IET CONTROL, ROBOTICS AND SENSORS SERIES 104

Control of Mechatronic Systems

Other volumes in this series:

Control of Mechatronic Systems

Levent Güvenç, Bilin Aksun Güvenç,
Burak Demirel and Mümin Tolga Emirler

The Institution of Engineering and Technology

Published by The Institution of Engineering and Technology, London, United Kingdom

The Institution of Engineering and Technology is registered as a Charity in England & Wales (no. 211014) and Scotland (no. SC038698).

© The Institution of Engineering and Technology 2017

First published 2017

The Institution of Engineering and Technology
Michael Faraday House
Six Hills Way, Stevenage
Herts, SG1 2AY, United Kingdom

www.theiet.org

British Library Cataloguing in Publication Data
A catalogue record for this product is available from the British Library

ISBN 978-1-78561-144-5 (hardback)
ISBN 978-1-78561-145-2 (PDF)

Typeset in India by MPS Limited
Printed in the UK by CPI Group (UK) Ltd, Croydon

Levent Güvenç dedicates this book to his parents
Aysel Güvenç and Ömer Güvenç

Bilin Aksun-Güvenç dedicates this book to her parents
Tanju Aksun and Alparslan Aksun

Levent Güvenç and Bilin Aksun-Güvenç dedicate this book to their son
Kunter Güvenç

Burak Demirel dedicates this book to his wife Esther Meijer

Mümin Tolga Emirler dedicates this book to his parents
Zehra Emirler and Raşit Emirler, and to his brother Muammer Mert Emirler

Contents

Preface

This book has evolved from the doctoral dissertation work of the second author, Prof. Bilin Aksun-Güvenç. Prof. Levent Güvenç and Prof. Bilin Aksun-Güvenç have taught the material in this book as a graduate level course for many years at Istanbul Technical University, and later at Istanbul Okan University. Most recently during 2016, Dr. Burak Demirel has taught the material in the book at the University of Paderborn, and Prof. Levent Güvenç has taught the material in the book at the Ohio State University. The audience for these courses were graduate students in Mechatronics Engineering, Mechanical Engineering, Electrical Engineering, Aerospace Engineering and Control Engineering who had taken at least one, preferably two courses on control systems analysis and design, including some exposure to discrete-time systems.

All authors have used the material in this book in their research work for successfully designing robust controllers for mechatronic systems from various disciplines. This book is intended for early-stage doctoral students who wish to get exposed to control methods that have been field-tested in a wide variety of mechatronic applications. The authors also believe that the book will be useful for practicing engineers who design and implement feedback control systems. The material in this book can be taught in a fourteen-week one-semester long course, with several project type assignments so that the graduate students can also master the application of the methods learned to practical problems.

The authors' graduate students and research collaborators have contributed to several topics covered in the book through joint publications, and their names may be found in the references cited at the end of each chapter.

About the authors

Levent Güvenç is a professor of mechanical and aerospace engineering at the Ohio State University with a joint appointment at the Electrical and Computer Engineering Department and conducts his research in the Center for Automotive Research (CAR), USA. He received the Ph.D. degree in Mechanical Engineering from the Ohio State University in 1992. He is recognized for his significant contributions in applied robust control, mechatronics, cooperative mobility of road vehicles, automotive control, and control applications in atomic force microscopy. He is a member of the International Federation of Automatic Control (IFAC) Technical Committees on Automotive Control, Mechatronics and Intelligent Autonomous Vehicles, and the IEEE Technical Committees on Automotive Control and Intelligent Vehicular Systems and Control. He served as department chair of mechanical engineering at Istanbul Okan University from 2011 to 2014, and was the founder and director of Mekar Mechatronics Research Labs and coordinator of team Mekar in the 2011 Grand Cooperative Driving Challenge. He is the founder and director of the Automated Driving Lab at the Ohio State University. He is an ASME fellow.

Bilin Aksun-Güvenç is currently a visiting professor in the Department of Mechanical and Aerospace Engineering and the Center for Automotive Research (CAR) of the Ohio State University, USA. She previously worked in Istanbul Technical University and Istanbul Okan University as a professor of mechanical engineering. She received the Ph.D. degree in 2001 from Istanbul Technical University. Her expertise is in automotive control systems – primarily vehicle dynamics controllers, such as electronic stability control, adaptive cruise control, cooperative adaptive cruise control, and collision warning and avoidance systems – and autonomous vehicles, intelligent transportation systems (ITS) and smart cities. Her current research focuses on ITS, on-demand automated shuttles for elderly, automated shuttles for a smart city. She was a member of team Mekar in the 2011 Grand Cooperative Driving Challenge. She is a member of the International Federation of Automatic Control (IFAC) Technical Committees on Automotive Control and on Mechatronics, and worked as a project evaluator and reviewer for European Union Framework Research Programs. She is a founding member of the Automated Driving Lab at the Ohio State University.

Burak Demirel received his B.Sc. degree both in Mechanical Engineering in 2007 with Summa Cum Laude and in Control Engineering (double major) in 2009 with high honour, and also received his M.Sc. degree in Mechatronics Engineering in 2009 from Istanbul Technical University, Turkey. He completed in 2015 his Ph.D. studies in the Department of Automatic Control of the School of Electrical Engineering at the KTH Royal Institute of Technology, Sweden. He is currently a postdoctoral researcher in Chair for Automatic Control at Paderborn University. His current research interests include cyber-physical systems, networked control systems, distributed algorithms, and mechatronic systems.

Mümin Tolga Emirler is currently an assistant professor in the Department of Mechanical Engineering of Istanbul Okan University, Turkey. He received his B.Sc. degree both in Mechanical Engineering in 2007 and in Manufacturing Engineering (double major with high honour) in 2008, his M.Sc. degree in Mechatronics Engineering in 2010 and his Ph.D. degree in Mechanical Engineering in 2015, all from Istanbul Technical University, Turkey. He worked as a postdoctoral scholar at the Ohio State University Center for Automotive Research in 2015. His current research interests include applied robust control systems, vehicle dynamics control, automotive control systems, autonomous vehicles and mechatronic systems. He is a junior member of the International Federation of Automatic Control (IFAC) Technical Committee on Automotive Control.

Control of mechatronic systems software

All examples in this book were prepared using MATLAB® m-files for computation and Simulink® for simulations. The MATLAB m-files were prepared and presented in the earlier work of the authors in the form of a MATLAB Graphical User Interface (GUI) that was named COMES. More information on the COMES toolbox can be found in the website:

http://www.mekarlab.net/comes.html

The Conventional Control part of COMES toolbox, which works in the latest version of MATLAB (2016b), was used for the computations in Chapter 3 of this book. The other portions of COMES toolbox on parameter space analysis and design, disturbance observer and repetitive control compensation using mixed sensitivity bound parameter space computations, work in earlier versions of MATLAB.

Instead of providing one MATLAB GUI like COMES and trying to update it continuously for changes that take place in MATLAB, the authors have decided to provide the m-files used in the computations in the course website. This is the approach that the authors are now using while teaching courses based on the book. A main program and its subprograms like the RobustControlDesign.m command line interactive MATLAB program and its subprograms are used in solving selected examples and case studies in each chapter. All such m-files can be found at the course software web page at

http://mekar.osu.edu/comes

on a chapter-by-chapter basis. The authors believe that this is a better approach for providing and updating software.

Chapter 1

Introduction

We begin the book with an introduction to mechatronic systems, the need for controlling them and the control methods that will be presented in this book for this purpose. Today's engineers face truly mechatronic plants and real control design challenges. They are required to do multi-domain modelling and use well-established control methods that have been successfully applied earlier for controlling mechatronic systems. This book concentrates on several of these control methods that have been successfully applied to real-world problems and are embedded in a lot of the products that we use in our everyday life.

1.1 Introduction and background

There are many different definitions of the term *mechatronics* that are available in the literature. Common to most of these definitions is *mechatronics* that is the synergistic combination of mechanical systems, electronics, control systems and computers as illustrated in Figure 1.1. The name MECHATRONICS was created by combining the two words: MECHAnics and ElecTRONICS by the Japanese engineer Tetsuro Mori in the trademark application of his company Yaskawa Electric Corporation; see, e.g., [1]. The company later gave up this trademark and mechatronics is now a widely used and accepted term. Mechatronics is taught as a course in many institutions. There are also departments of mechatronics in many universities. There are many textbooks, reference books and archival journals dedicated to mechatronics.

This synergistic approach to the design of systems in a wide variety of domains (mechanical, electrical, electromechanical, fluid and thermal) has resulted in the mechatronic design approach. The resulting systems are called mechatronic systems in this book. We are all surrounded by mechatronic systems around us. An industrial robot is a classic example of a mechatronic system including all aspects of mechanics, electronics and computing inherently in its design. Mechatronic systems play a significant role in many different application areas including automotive, aerospace, medical, materials processing, manufacturing, defence, chemical systems and consumer products. A passenger car like the one shown in Figure 1.2 with its integration of mechanical suspension and steering systems with the thermal engine and fluid fuel system, electronic sensors/actuators and electronic control units for computer control is an excellent example of a highly coupled, complicated mechatronic system.

This book concentrates on control methods and controller design techniques that have been successfully applied in mechatronics applications. The aim is to introduce graduate students with a basic control systems background to an array of control techniques, which they can easily implement, and use to meet the required performance

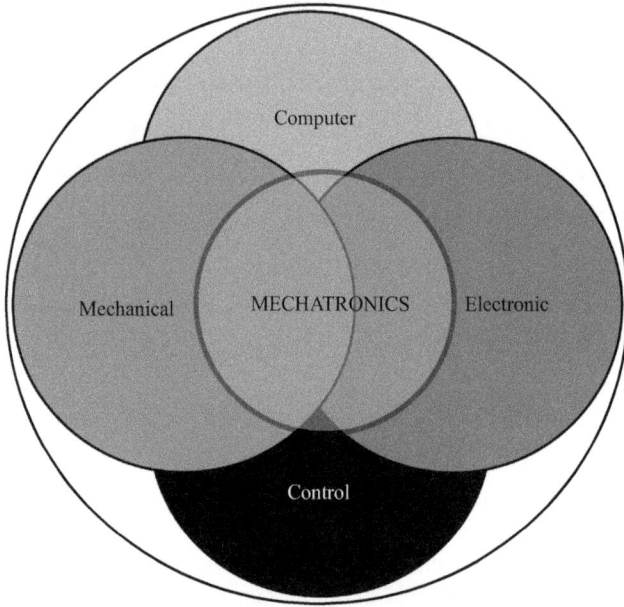

Figure 1.1   Venn diagram illustrating the synergy of mechatronics

Figure 1.2   A passenger car as an example of a complicated mechatronic system

specifications for their mechatronic application. The parameter space design method is used to obtain robust controllers. The controller architectures treated comprise of conventional control (phase lead, lag and PID control), input shaping, repetitive control and disturbance observer (model regulator) control. Some of these control methods and design approaches have not yet been published in books even though there is an abundant literature on them in the form of technical papers reporting successful applications and presenting details of the design approach used. The control architectures introduced in this book are presented in a holistic approach to solving real mechatronics problems. Parameter space methods are presented and used for achieving robust controllers. Examples in the form of case studies from the motion control, automotive control and atomic force microscopy (AFM) control domains are used in the book to further illustrate the topics covered. The model of a lab-type experimental rotary motion control system is used in several examples to illustrate the methods in the book [2].

1.2 Overall control architecture

The overall control architecture presented in this book is illustrated in Figure 1.3. All of the individual control methods introduced in the subsequent chapters of this book are shown in the same block diagram in Figure 1.3. However, it is not necessary to use all of these control methods at the same time. They are all shown in the same figure for illustrative purposes. The disturbance observer (model regulator) is used to reject disturbances and also regulate an uncertain plant close to its nominal model. A conventional controller is then designed for this nominal model and implemented. In the case of periodic disturbances and/or reference inputs, a repetitive controller is added within the feedback loop to reject periodic components of the disturbance and/or to follow periodic components of the reference input more accurately. Finally, an approximate inverse of the feedback-controlled system is used as a preview controller

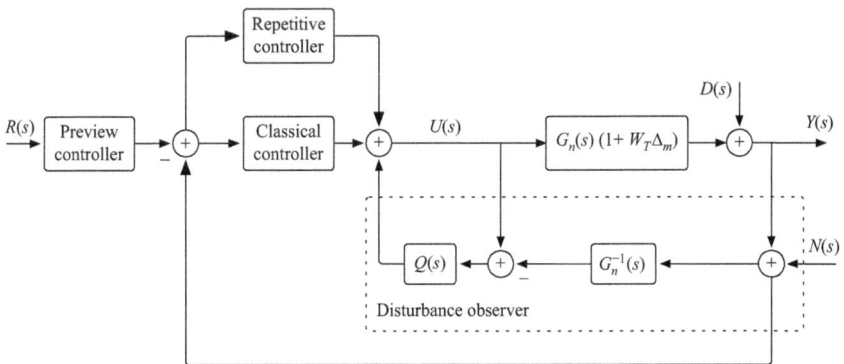

Figure 1.3   Overall mechatronic systems control architecture

to shape the known input (or disturbance) as an input-shaping filter. The approximate inverses are also used within the disturbance observer and repetitive controller filters.

1.3   Overview of control methods

Parameter space based multi-objective solutions will be formulated and used in all the chapters except for Chapters 3 and 4 as the main design method. The parameter space method for robust control is, therefore, presented in Chapter 2 as a pre-requisite to the other chapters. Real root boundary (RRB), complex root boundary (CRB) and infinite root boundary (IRB) computations are presented and applied to Hurwitz stability region and $\mathscr{D}$-stability region determination in parameter space. The physical significance of the $\mathscr{D}$-stability region boundary is explained. The method of mapping frequency domain bounds to parameter space is presented. The frequency domain bounds considered are mixed sensitivity bounds and phase margin bounds. Automated path following of a road vehicle and a Qube servo application are the case studies illustrating the parameter space control design tools.

Chapter 3 is about the conventional control methods, and phase lead, phase lag, phase lag-lead, PD, PI and PID controllers are dealt with. Design for a specified bandwidth and error constant is presented. The design of conventional controllers using parameter space methods is also presented. Chapter 3 also has a section on the use of optimisation-based MATLAB®/Simulink® PID tuners. The chapter ends with design examples on lane keeping assistance of a road vehicle and a Qube servo conventional controller design example.

Chapter 4 is about input shaping which is also known as preview control. A discrete-time closed-loop control system is considered, and its approximate inverse filter that is not causal is designed to achieve desirable input-to-output behaviour. The input is assumed to be known in advance. Thus, the approximate inverse filter, also known as the preview filter as it is not causal, is applied to the known input offline, resulting in a new input which is applied online to the closed loop system (hence the name input shaping). This chapter starts with a characterisation of non-minimum phase zeros and approximate inversion methods used to treat them, as they cannot be inverted directly. The zero phase (ZP) error, ZP with gain compensation (ZPG), ZP with extended gain compensation (ZPGE), ZP with optimal gain compensation (ZPGO) and truncated series approximation (TSA) methods are presented. The examples in Chapter 4 use control of a hydraulic actuator and an electrohydraulic material testing machine as the plants under control.

The disturbance observer (also known as the model regulator) is a well-known two-degrees-of-freedom control architecture with excellent disturbance rejection and model regulation properties. Chapter 5 introduces both continuous-time and discrete-time disturbance observer control systems. Parameter space based disturbance observer control is presented as the main analytical design method. The Communication Disturbance Observer (also called the Communication Model Regulator) is introduced to handle plants with time delay. A MIMO decoupling type extension of the model regulator is given. The chapter has several design examples.

The first example is an application example using the Qube servo. The second example is the use of the SISO disturbance observer for yaw stability control of a road vehicle. The third example is on MIMO disturbance observer control for decoupling the two axes of motion in a piezo-tube actuator used in an AFM. The fourth example is on MIMO disturbance observer control for a four-wheel steering vehicle. The fifth example is on the application of the Communication Disturbance Observer to vehicle yaw stability control over the CAN bus.

Chapter 6 is dedicated to repetitive control. Repetitive control uses a time delay element in a feedback loop to reduce tracking error for systems with a periodic reference or disturbance input of known period. The SISO continuous-time and discrete-time repetitive control architectures and design methods based on minimising mixed sensitivity at the fundamental periodic frequency, and its harmonics are introduced. Chapter 6 has two design examples. The first example is on an AFM application. The second example is a Qube servo application.

The book ends with a summary of the results of the previous chapters in the last chapter along with conclusions. How to access the MATLAB m-files used and the COMES (COntrol of MEchatronic Systems) toolbox for MATLAB is presented in "Control of mechatronic systems software" section. Appendix A focuses on rapid controller prototyping and hardware-in-the-loop simulation for implementing and testing the control methods presented in the book.

1.4 Software

The widespread use of interactive, computer-aided tools in the last three decades is a very significant development in control engineering made possible by cheaper and faster computing equipment [3]. Jia and Schaufelberger [4] proposed many interesting ideas about the use of computer-aided education in the field of automatic control. Johansson *et al.* [5] developed an interactive design tool, which aims at designing classical controllers, such as lead, lag or lead–lag compensators, for use in automatic control education. Azemi and Yaz [6] gathered all programs that are used in teaching their optimal control course and prepared a toolbox for MATLAB to design optimal controllers. Instead of using computer-aided control system design in an educational manner, it can also be utilised to solve real engineering problems. Sienel *et al.* [7] created an interactive MATLAB-based robust control toolbox called Paradise to design and analyse control systems by using symbolic parameter space methods for robust control. Similarly, Sakabe *et al.* [8] developed a MATLAB-based toolbox to design robust controllers based on the parameter space approach. Hyodo *et al.* [9] also improved this MATLAB-based toolbox for robust parametric control. In reference [10], a robust control toolbox, which combines the MATLAB Robust Control Toolbox, LMI, and μ-Analysis, and Synthesis Toolbox, is given. Vivero and Liceago-Castro [11] introduced novel software in MATLAB to analyse and design multivariable control systems. Boyle *et al.* [12] designed an interactive design program in MATLAB for the frequency domain analysis and design of multivariable feedback systems. Campa *et al.* [13] created a multivariable design program for linear system analysis and robust control synthesis.

In a manner similar to all the references as mentioned earlier, most of the design approaches presented in this book were also coded in MATLAB during the period 2000 to 2010. They were compiled together in the form of a GUI (Graphical User Interface) in MATLAB with the name COMES (COntrol of MEchatronic Systems toolbox) [14]. The COMES toolbox was presented in several papers and presentations by the authors and their colleagues [14, 15]. The current version of COMES works with an old version of MATLAB (2007). As it is hard to continuously update the COMES GUI, the m-files that constitute it are made available in a web site for this book for free download and use. The m-files for some of the textbook examples in this book are also made available on the same site. Section "Control of mechatronic systems software" contains information on how these m-files can be accessed.

References

[1]   T. Mori, "Mechatronics," Yasakawa Internal Trademark Application, Memo 21.131.01, July 12 1969.

[2]   (2017, Jan.). [Online]. Available: http://www.quanser.com/courseware/qubeservo_matlab/

[3]   B. Aksun-Güvenç, "Applied robust motion control," Ph.D. dissertation, Istanbul Technical University, Istanbul, Turkey, 2001.

[4]   L. Jia and W. Schaufelberger, *Software for Control Engineering Education.* Zürich: Verlag der Fachvereine an der ETH-Zürich, 1995.

[5]   M. Johansson, M. Gäfvert, and K. J. Åström, "Interactive tools for education in automatic control," *IEEE Control Systems*, vol. 18, no. 3, pp. 33–40, June 1998.

[6]   A. Azemi and E. E. Yaz, "Using MATLAB in a graduate electrical engineering optimal control course," in *Proceedings of IEEE Frontiers in Education Conference*, 1997.

[7]   W. Sienel, T. Bünte, and J. Ackermann, "PARADISE – Parametric robust analysis and design interactive software environment: A MATLAB-based robust control toolbox," in *Proceedings of IEEE International Symposium on Computer-Aided Control System Design*, 1996.

[8]   K. Sakabe, H. Yanami, H. Anai, and S. Hara, "A MATLAB toolbox for robust control synthesis by symbolic computation," in *Proceedings of SICE Annual Conference*, 2004.

[9]   N. Hyodo, M. Hong, H. Yanami, H. Anai, and S. Hara, "Development of a MATLAB toolbox for parametric robust control – algorithms and functions," in *Proceedings of International Joint Conference SICE-ICASE*, 2006.

[10] G. J. Balas, A. Packard, M. G. Safonov, and R. Y. Chiang, "Next generation of tools for robust control," in *Proceedings of the American Control Conference*, 2004.

[11]   O. Vivero and J. Liceago-Castro, "MIMO toolbox for MATLAB," in *Proceedings of Annual IEEE Student Paper Conference*, 2008.

[12]   J. M. Boyle, M. P. Ford, and J. M. Maciejowski, "Multivariable toolbox for use with MATLAB," *IEEE Control System Magazine*, vol. 9, no. 1, pp. 59–65, Jan. 1989.

[13]   G. Campa, M. Davini, and M. Innocenti, "MvTools: Multivariable Systems Toolbox," in *Proceedings of IEEE International Symposium on Computer-Aided Control System Design*, 2000.

[14]   B. Demirel, "Interactive computer-aided controller design for mechatronic systems," M. Sc. Thesis, Istanbul Technical University, Istanbul, Turkey, Sept. 2009.

[15]   B. Demirel and L. Güvenç, "Control of mechatronic systems – COMES toolbox," in *ASME Engineering Systems Design and Analysis Conference*, 2010.

Chapter 2

Parameter space based robust control methods

The parameter space approach to robust control is presented in this chapter as a pre-requisite to the remaining chapters. The well-established method of mapping the left-half plane for Hurwitz stability or a smaller bounded subregion of the left-half plane for $\mathscr{D}$-stability are presented first. Mapping of phase and gain margin bounds and different sensitivity bounds to parameter space are introduced later in the chapter with illustrative examples.

2.1 Introduction

The well-established conventional approach to controller design is based on a single nominal plant model. This model is usually an experimentally validated model that is only as accurate as is needed by the specific control application. In practice, the model that is never used exactly matches the true behaviour of the plant, especially at high frequencies where signal-to-noise ratio becomes small and experimental identification is difficult. The difference between the nominal model and the true plant is called model uncertainty and is caused by plant model parameter variations or neglected high-frequency dynamics. The compensation of model uncertainty is one of the main reasons for using feedback along with stabilisation and disturbance rejection. A family of models instead of just the nominal plant model characterises a plant model with uncertainty. Controllers are implemented either digitally or using a dedicated hardware like an electronic or a hydraulic circuit. As a consequence, controller parameters may also have a relatively smaller amount of uncertainty associated with them in practice. Stability and performance should not be affected by these small possible errors in controller parameters. Otherwise, the result will be a fragile controller according to [1]. The parameter space approach to controller design that is treated in this chapter naturally results in non-fragile controllers.

Since model uncertainty cannot be avoided, robust controllers should be designed and used so that stability and a desired level of performance is achieved for each element of the family of possible uncertain plants. There are several methods for robustness analysis and robust controller design. Two major approaches are the parametric and the frequency domain approaches to robust control. On the frequency domain side, there are the well-known $\mathscr{H}_\infty$ optimisation methods, e.g., [2–4], which treat unstructured uncertainty and are developed mainly for handling MIMO systems.

Structured uncertainty is treated less conservatively in the frequency domain approach by the structured singular value (abbreviated henceforth as SSV) method [5].

A very significant development in the parametric approach was provided by Kharitonov's Theorem [6] for uncertain characteristic polynomials with unrelated interval coefficients and later by the Edge Theorem [7] for affine coefficients in the uncertain polynomials. Out of the parametric methods of robust control, the parameter space method is concentrated upon in this book. This chapter introduces the parameter space method for robustness analysis and robust controller design.

2.2   Hurwitz stability

The parameter space approach can be used to determine a set of coefficients for a given controller structure, which simultaneously stabilises a finite number of plants. We first define the plant model uncertain parameter set as $q = [q_1, q_2, \ldots, q_n]^\mathsf{T} \in \mathbb{Q}$. Note that $\mathbb{Q}$ is the family of all possible uncertain plants for which we want the characteristic polynomial to be Hurwitz-stable. Similarly, we also define the set of all controller parameters that can be designed or tuned as $k = [k_1, k_2, \ldots, k_n]^\mathsf{T} \in \mathbb{K}$. Notice that $\mathbb{K}$ is the family of all possible controller parameter combinations. The use of uncertain plant parameters q corresponds to robustness analysis of an existing controller while the use of uncertain controller parameters k corresponds to robust controller design or tuning controller gains for robustness. A combination of plant and controller parameters – q and k – is useful for gain scheduling [8]. A survey of parameter space methods can be found in [9].

In the algebraic approach to the parameter space method, a direct correlation between roots of the characteristic equation, $p(s, q, k) = 0$, and the uncertain parameters or controller gains appearing in it are determined in a manner suitable for a graphical display in parameter space; see, e.g., [10–12]. Although this method is more appropriate for SISO systems, the characteristic equation can be computed and used similarly for MIMO systems as well.

Consider the feedback control system shown in Figure 2.1. Its loop gain is defined by

$$L(s) \triangleq C(s)G(s)H(s), \tag{2.1}$$

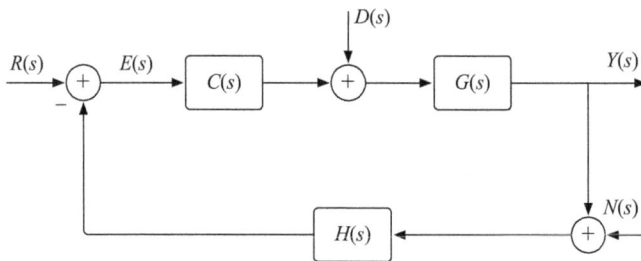

Figure 2.1   Feedback controlled system

where $C(s)$, $G(s)$ and $H(s)$ are the controller, plant and sensor dynamics models, respectively. The sensitivity closed-loop transfer function is given by

$$S(s) \triangleq \frac{E(s)}{R(s)} = \frac{1}{1 + L(s)} = \frac{1}{1 + C(s)G(s)H(s)}, \tag{2.2}$$

where $E(s)$ is the error and $R(s)$ is the input in the Laplace domain. The characteristic equation of the feedback-controlled system, illustrated in Figure 2.1, is obtained by setting the denominator polynomial of (2.2) equal to zero as

$$p(s) = \text{num}\{1 + L(s)\} = \text{num}\{1 + C(s)G(s)H(s)\}. \tag{2.3}$$

Expressing all transfer functions in terms of their numerators and denominators results in

$$C(s) = \frac{N_C(s)}{D_C(s)}, \quad G(s) = \frac{N_G(s)}{D_G(s)}, \quad H(s) = \frac{N_H(s)}{D_H(s)},$$

and

$$p(s) = \text{num}\left\{1 + \frac{N_C(s)}{D_C(s)}\frac{N_G(s)}{D_G(s)}\frac{N_H(s)}{D_H(s)}\right\}$$

$$= D_C(s)D_G(s)D_H(s) + N_C(s)N_G(s)N_H(s) = 0.$$

Considering plant model and controller parameter uncertainties, we can rewrite the characteristic equation as

$$p(s, \boldsymbol{q}, \boldsymbol{k}) = a_{n+m}(\boldsymbol{q}, \boldsymbol{k})s^{n+m} + a_{n+m-1}(\boldsymbol{q}, \boldsymbol{k})s^{n+m-1}$$

$$+ \cdots + a_1(\boldsymbol{q}, \boldsymbol{k})s + a_0(\boldsymbol{q}, \boldsymbol{k}), \tag{2.4}$$

which is an $n + m$ degree polynomial with uncertain coefficients. Notice that n is the degree of the plant while m is the degree of the controller. Assume that the sensor dynamics, $H(s)$, does not contain any uncertainty and tuneable parameters.

Hurwitz stability requires all roots of the characteristic equation, $p(s, \boldsymbol{q}, \boldsymbol{k}) = 0$, to lie in the left-half plane as illustrated in Figure 2.2. According to the Boundary

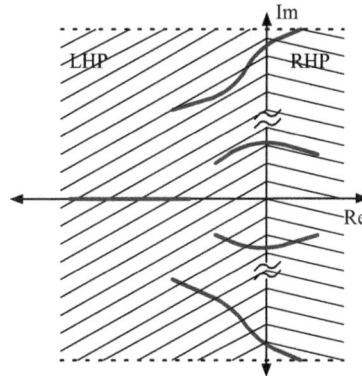

Figure 2.2   Hurwitz stability region and boundary crossing

Crossing Theorem [13], which is also called the zero exclusion principle in the literature, roots cannot just jump from the left-half to the right-half plane without crossing the imaginary axis boundary in between. The *Boundary Crossing Theorem* is expressed more analytically as follows:

The family of polynomials $p(s, q, k)$ for any $q \in \mathbb{Q}$ and $k \in \mathbb{K}$ is robustly stable, if and only if

(a)   there exists at least one stable polynomial $p(s, q, k)$ for any $q \in \mathbb{Q}$ and $k \in \mathbb{K}$, and

(b) $j\omega \notin$ roots$\{p(s, q, k)\}$ with $q \in \mathbb{Q}$ and $k \in \mathbb{K}$ for all $\omega \geq 0$.

Starting with a Hurwitz stable characteristic polynomial, Hurwitz stability is maintained as long as the roots of $p(s, q, k)$ do not cross the imaginary axis due to uncertainty in the plant model $q \in \mathbb{Q}$ and changes in the controller coefficients $k \in \mathbb{K}$. If we solve for and plot the roots of $p(j\omega, q, k) = 0$ as parameterised by frequency ω, we will obtain the boundary between the stable and unstable regions in the chosen parameter space. Since $p(j\omega, q, k) = 0$ is a complex equation, it results in at most two independent equations as

$$\text{Re}\{p(j\omega, q, k)\} = 0, \tag{2.5}$$

$$\text{Im}\{p(j\omega, q, k)\} = 0, \tag{2.6}$$

which can be solved for two chosen parameters in the uncertain sets $q \in \mathbb{Q}$ and $k \in \mathbb{K}$. As shown in Figure 2.2, the roots of $p(s, q, k)$ can cross the imaginary axis in three different ways. The first one is a pair of complex conjugate roots crossing the imaginary axis and is, therefore, called a CRB. The characteristic equation (2.4) and the CRB equations (2.5) and (2.6) result in

$$\text{Re}\{p(j\omega, q, k)\} = a_0(q, k) - a_2(q, k)\omega^2 + a_4(q, k)\omega^4 - \cdots = 0, \tag{2.7}$$

$$\text{Im}\{p(j\omega, q, k)\} = a_1(q, k)\omega - a_3(q, k)\omega^3 + a_5(q, k)\omega^5 - \cdots = 0, \tag{2.8}$$

which can be solved for two chosen parameters in $q \in \mathbb{Q}$ and $k \in \mathbb{K}$ in terms of frequency ω and plotted in the parameter space of the chosen parameters as shown in Figure 2.3. The solution of (2.7) and (2.8) for each frequency ω results in one point in the parameter space plot shown in Figure 2.3. It is common to use a gridding approach to grid frequency ω and solve (2.7) and (2.8) for each frequency, obtaining an open line/curve or closed curve in parameter space. The line or curve corresponds to the CRB in the complex plane. With a closed curve, either the inside or the outside is a solution, meaning achievement of Hurwitz stability. With an open line/curve either one side or the other is a solution and the other side does not satisfy Hurwitz stability. Which region corresponds to a solution in the parameter space is checked interactively in the available MATLAB® code for parameter space analysis and design including the COMES toolbox. Solutions can be obtained numerically or symbolically. Use of symbolic manipulation makes it easier to code the required computations.

The other two cases of crossing the imaginary axis are called the RRB and IRB as shown in Figure 2.2 and correspond to degenerate cases in which one of the equations (2.5), (2.6) vanish and we are left with only one equation to determine the

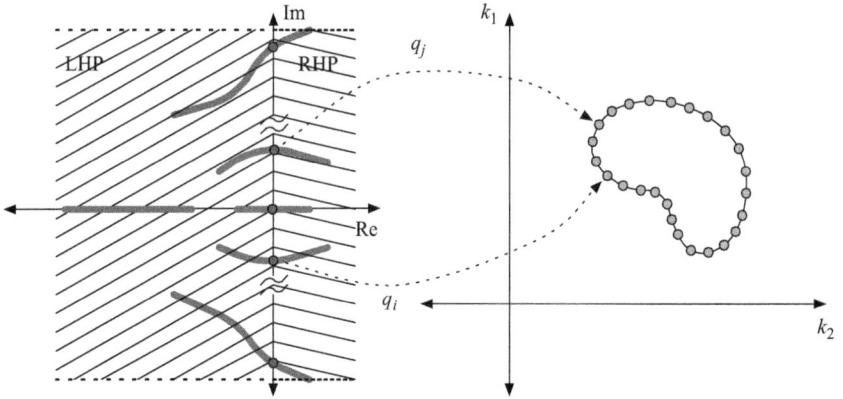

Figure 2.3   Mapping Hurwitz stability CRB to parameter space

two chosen parameters in $q \in \mathbb{Q}$ and $k \in \mathbb{K}$, shown as q_i and q_j in Figure 2.3. While the CRB solution for two chosen parameters corresponds to a point in the parameter space for each ω, the RRB or IRB solutions or degenerate cases, if they exist, will result in curves or lines in the parameter space for each ω.

Let us first consider the RRB. The imaginary axis is crossed at the real root $s = j\omega = 0$ in this case at $\omega = 0$. The characteristic equation (2.4) becomes

$$p(0, q, k) = a_0(q, k) = 0, \tag{2.9}$$

which is a single equation as the characteristic polynomial does not have an imaginary part in this case. The RRB solution is obtained by solving (2.9) for the two chosen parameters in $q \in \mathbb{Q}$ and $k \in \mathbb{K}$. The line/curve obtained in parameter space corresponds to $\omega = 0$.

Let us finally consider the IRB which corresponds to frequency ω reaching infinity, i.e., boundary crossing at the top or bottom of the imaginary axis. This is also a degenerate case and leads to

$$\lim_{\omega \to \infty} p(j\omega, q, k) = \lim_{\omega \to \infty} (a_{n+m}(q, k)(j\omega)^{n+m} + \cdots + a_0(q, k)(j\omega) + a_0(q, k)) = 0,$$
$$\tag{2.10}$$

which results in

$$a_{n+m}(q, k) = 0, \tag{2.11}$$

as the equation to be solved for the IRB. Either the real or imaginary part of the characteristic polynomial in (2.10) will be identically zero based on whether $n + m$ is odd or even. The IRB solution is obtained by solving (2.11) for the two chosen parameters in $q \in \mathbb{Q}$ and $k \in \mathbb{K}$. The line/curve obtained in parameter space corresponds to $\omega \to \infty$. The reader should refer to [12] for more detailed information on the parameter space approach including the topic of singular frequencies which is only treated briefly in Section 2.7 in this book.

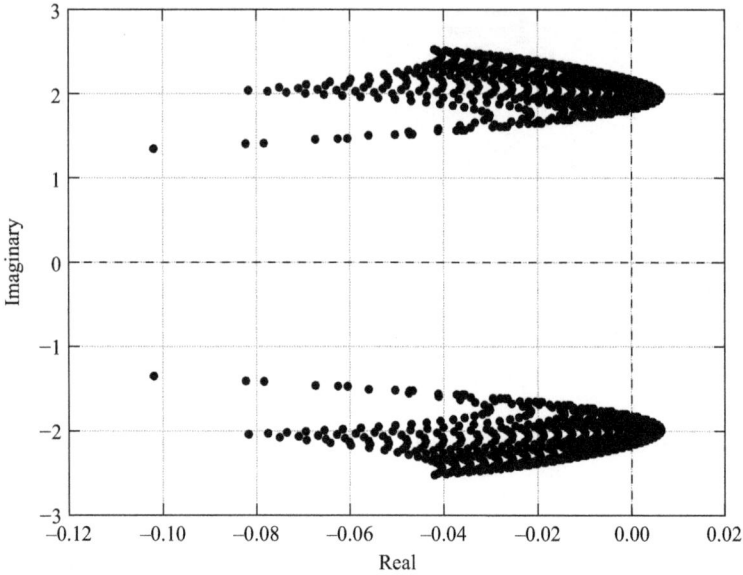

Figure 2.4   Variation of characteristic equation roots due to uncertain parameters

Example 2.1 (Uncertainty in characteristic polynomial)

This example considers the effect of two uncertain parameters, i.e., q_1 and q_2, on the closed-loop pole locations. The family of characteristic equations with uncertainty is given by

$$\mathscr{P}_q = \{p(s, q_1, q_2) \mid 0 \le q_1 \le 2,\ 0.25 \le q_2 \le 3\}, \tag{2.12}$$

where

$$\begin{aligned} p(s, q_1, q_2) &= s^3 + (2 + q_1 + q_2)s^2 + (2 + q_1 + q_2)s \\ &\quad + 6(q_1 + q_2) + 2q_1 q_2 + 2.25. \end{aligned} \tag{2.13}$$

The uncertain parameter limits in (2.12) are gridded to plot the variation of the roots of the characteristic equation in (2.13). The loci of characteristic equation roots that is obtained is illustrated in Figure 2.4. There are three characteristic equation roots for each possible combination of the uncertain parameters q_1 and q_2. One of these is real and the other two are a complex conjugate pair. The real root is always stable and only the complex conjugate pair is shown in Figure 2.4. It is clear that some choices of the uncertain parameters result in Hurwitz instability.

Example 2.2 (Computation of Hurwitz stability region)

This example is on Hurwitz stability of the proportional-plus-integral (PI) controlled second-order plant

$$G(s) = \frac{1}{s^2 + 7s + 10}. \tag{2.14}$$

The uncertain PI controller is given by

$$C(s) = \frac{k_1 s + k_2}{s}, \tag{2.15}$$

and $H(s) = 1$. This is a controller design or tuning example. The characteristic equation becomes

$$1 + C(s)G(s)H(s) = 1 + \frac{k_1 s + k_2}{s} \frac{1}{s^2 + 7s + 10}, \tag{2.16}$$

which becomes

$$s(s^2 + 7s + 10) + k_1 s + k_2 = s^3 + 7s^2 + (10 + k_1) + k_2 = 0. \tag{2.17}$$

The CRB equation with $s = j\omega$ and $0 < \omega < \infty$ becomes

$$p(j\omega, k_1, k_2) = -j\omega^3 - 7\omega^2 + (10 + k_1)j\omega + k_2 = 0, \tag{2.18}$$

which results in the two equations:

$$\text{Re}\{p(j\omega, k_1, k_2)\} = -7\omega^2 + k_2 = 0, \tag{2.19}$$

$$\text{Im}\{p(j\omega, k_1, k_2)\} = -\omega^3 + (10 + k_1)\omega = 0. \tag{2.20}$$

The two unknown parameters in (2.19) and (2.20) are k_1 and k_2 while the free parameter is ω. The numerical solution approach will be to grid the frequency ω from zero to a large value and solve (2.19) and (2.20) for k_1 and k_2. These two equations are very simple to solve in terms of frequency ω as

$$k_1 = \omega^2 - 10,$$
$$k_2 = 7\omega^2,$$

which result in

$$k_1 = \frac{k_2}{7} - 10. \tag{2.21}$$

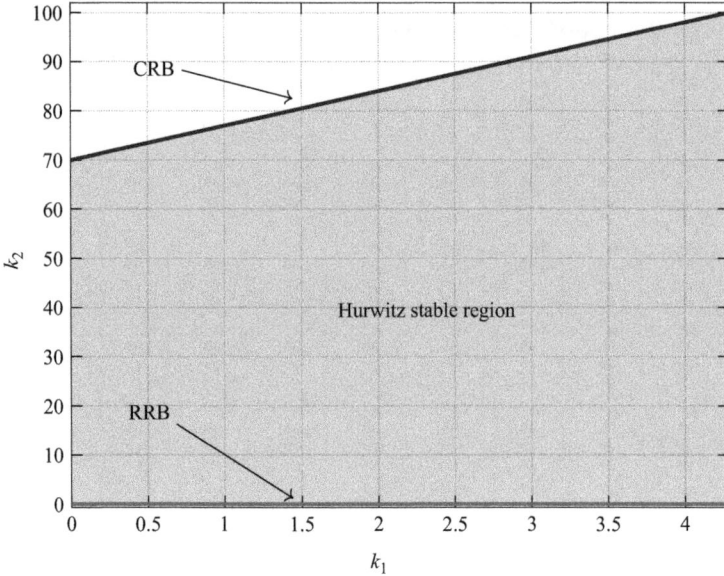

Figure 2.5   Parameter space region for Example 2.2

It is worth noting that (2.21) is an equation of line in the k_1 and k_2 parameter plane. The equation for the RRB at $s = j\omega = 0$ is

$$p(0, k_1, k_2) = k_2 = 0, \tag{2.22}$$

which is a horizontal straight line or the k_1-axis in the $k_1 - k_2$ plane. The equation of the IRB as $s = j\omega$ goes to $j\infty$ is

$$a_{n+m} = a_3 = 1 \neq 0, \tag{2.23}$$

which cannot be satisfied. The IRB does not provide a useful equation and does not have an effect on the $k_1 - k_2$ parameter plane which is influenced by the CRB and RRB for this example. The resulting parameter space region for Hurwitz stability is displayed in Figure 2.5.

Example 2.3 (Uncertainty in plant)

For the previous example, let us also consider uncertainty in the plant parameters given by

$$G(s) = \frac{1}{s^2 + q_1 s + q_0}, \tag{2.24}$$

and the PI controller in (2.15) is used with $H(s) = 1$. The characteristic equation becomes

$$1 + C(s)G(s)H(s) = 1 + \frac{k_1 s + k_2}{s} \frac{1}{s^2 + q_1 s + q_0} = 0, \tag{2.25}$$

or equivalently,

$$p(s, \boldsymbol{q}, \boldsymbol{k}) = s^3 + q_1 s^2 + (q_0 + k_1)s + k_2 = 0. \tag{2.26}$$

For Hurwitz stability of this family of characteristic polynomials, the CRB is given by

$$p(j\omega, \boldsymbol{q}, \boldsymbol{k}) = -j\omega^3 - q_1\omega^2 + (q_0 + k_1)j\omega + k_2 = 0, \tag{2.27}$$

with $0 < \omega < \infty$ which results in the real and imaginary component equations:

$$\text{Re}\{p(j\omega, \boldsymbol{q}, \boldsymbol{k})\} = -q_1\omega^2 + k_2 = 0, \tag{2.28}$$

$$\text{Im}\{p(j\omega, \boldsymbol{q}, \boldsymbol{k})\} = -\omega^3 + (q_0 + k_1)\omega = 0. \tag{2.29}$$

We can choose any two of the unknown parameters q_0, q_1, k_1, k_2 and carry out a parameter space analysis. Let us first do an uncertainty analysis by considering plant model uncertainty q_0 and q_1. Equations (2.28) and (2.29) become

$$q_1 = \frac{k_2}{\omega^2}, \tag{2.30}$$

$$q_0 = \omega^2 - k_1. \tag{2.31}$$

It is assumed that k_1 and k_2 are known through controller design for the nominal plant. Equations (2.30) and (2.31) can be solved by gridding frequency ω and obtaining (q_0, q_1) pairs corresponding to each value of frequency. The resulting curve or line can then be plotted in the $q_1 - q_0$ parameter plane. For this example, it is easy to eliminate frequency ω in (2.30) and (2.31), and obtain

$$q_1 = \frac{k_2}{q_0 + k_1}, \tag{2.32}$$

which is the equation of a hyperbola in the $q_1 - q_0$ parameter plane. The RRB is given by

$$p(0, \boldsymbol{q}, \boldsymbol{k}) = k_2 = 0, \tag{2.33}$$

which is not useful for plant model uncertainty analysis. The RRB does not provide any useful information when the integral gain k_2 is not zero. The IRB is given by

$$a_{n+m} = a_3 = 1 \neq 0, \tag{2.34}$$

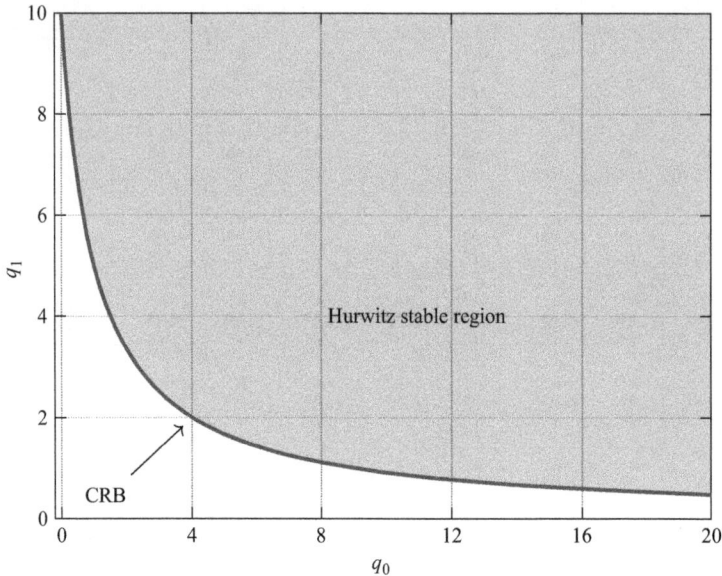

Figure 2.6   Parameter space region for Example 2.3

which does not introduce any useful information. Only the CRB is helpful to compute parameter space for the plant model uncertainty. The parameter space region that is obtained is shown for the choice of PI controller gains $k_1 = 1$ and $k_2 = 10$ in Figure 2.5. This set of PI controller gains resulted in Hurwitz stability for $q_1 = 7$ and $q_0 = 10$ in Example 2.1 according to the previous results in Figure 2.4. It is seen that the plant model parameters, i.e., $q_1 = 7$ and $q_0 = 10$, of Example 2.1 are in the stable region in Figure 2.5 as it should be.

Example 2.4 (Uncertainty in both plant and controller)

Let us also consider uncertainty in one plant model parameter q_0 and one controller gain k_2 in the previous example and work on a simple gain scheduled controller design for Hurwitz stability. We will assume that the uncertain plant model uncertainty is measurable. Equations (2.30) and (2.31) can be used to obtain

$$k_2 = q_1 q_0 + q_1 k_1 = 7q_0 + 7, \tag{2.35}$$

as the CRB equation where it is assumed that q_1 and k_1 are known. In (2.35), $q_1 = 7$ and $k_1 = 1$ were used. The RRB and IRB equations do not result in useful information. The CRB equation in (2.35) is a straight line as seen in the parameter space region in Figure 2.7.

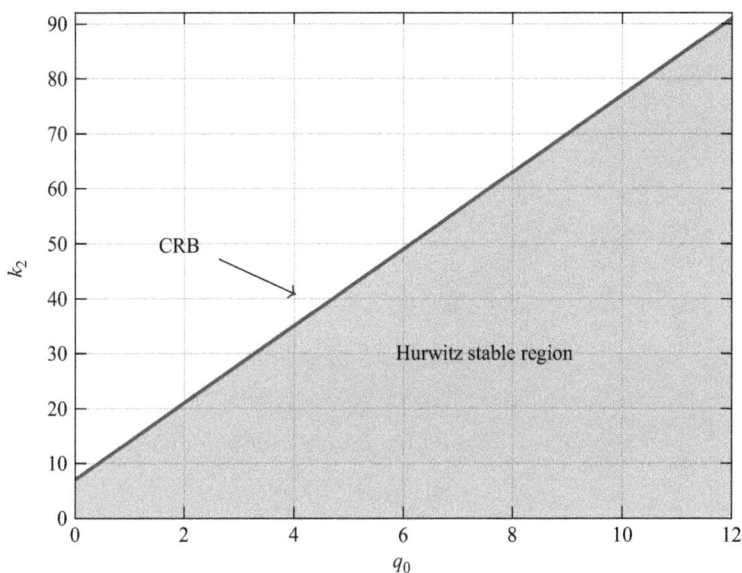

Figure 2.7 Parameter space region for Example 2.4

2.3 $\mathscr{D}$-stability

The procedure for computing the parameter space region, which ensures Hurwitz stability, can be readily extended to relative stability [14] (also known as $\mathscr{D}$-stability). $\mathscr{D}$-stability is a subset of Hurwitz stability, which requires all poles of the closed-loop system to lie within the unbounded left-half plane. A closed-loop system is called $\mathscr{D}$-stable if all poles of the closed-loop system belong to the predestined region $\mathscr{D}$ on the complex plane. $\mathscr{D}$-stability is, thus, a sufficient condition for Hurwitz stability which does not need to be checked for a $\mathscr{D}$-stable system. The $\mathscr{D}$-stability region and boundary used in this book is illustrated in Figure 2.8. The $\mathscr{D}$-stability region boundary ∂ consists of the three boundaries ∂_1, ∂_2 and ∂_3. The boundary ∂_1 corresponds to improved relative stability or a desired settling time as it specifies a maximum bound on the real part of poles which means an exponential decay that is faster than $e^{-\sigma t}$ where $-\sigma$ is the real axis intercept of ∂_1. For a dominant complex conjugate pole pair with real part $-\sigma$, the 98% settling time is given by $4/\sigma$.

The boundary crossing theorem is also applicable to $\mathscr{D}$-stability. The role of the imaginary axis is taken over by the $\mathscr{D}$-stability boundary in this case. We need to take a look at possible CRB, RRB and IRB solutions for each of the three boundaries ∂_1, ∂_2 and ∂_3 that poles must pass through before becoming $\mathscr{D}$-unstable. The boundary ∂_1 can only be crossed using the CRB and RRB. An IRB is not possible for ∂_1 as that boundary does not extend until infinity. The boundary ∂_2 mandates a minimum damping ratio such that $\zeta = \cos\theta$ in Figure 2.8. This value of ζ corresponds to a maximum allowable overshoot. The boundary ∂_3 mandates an upper bound on the

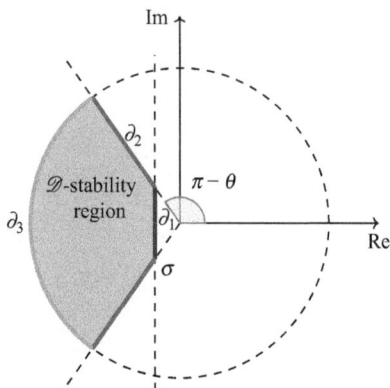

Figure 2.8 $\mathscr{D}$-stable region in the complex plane

natural frequency ω_n which is limited by R in Figure 2.8. In this book, ω_n will be used to approximate the bandwidth of the system. The boundary ∂_3 will be viewed as a limit on the actuator bandwidth. $\mathscr{D}$-stability in Figure 2.8 thus guarantees a desired settling or speed of response, an upper bound on allowable overshoot and a bound on the actuator bandwidth. There is only the CRB for boundary ∂_2, and CRB and RRB for boundary ∂_3.

Consider the boundary ∂_1. The CRB is computed by using

$$p(-\sigma +j\omega, \boldsymbol{q}, \boldsymbol{k}) = \sum_{j=0}^{n+m} a_j(\boldsymbol{q}, \boldsymbol{k})(-\sigma +j\omega)^j = 0, \qquad (2.36)$$

which results in the real and imaginary component equations:

$$\mathrm{Re}\{p(-\sigma +j\omega, \boldsymbol{q}, \boldsymbol{k})\} = 0, \qquad (2.37)$$

$$\mathrm{Im}\{p(-\sigma +j\omega, \boldsymbol{q}, \boldsymbol{k})\} = 0. \qquad (2.38)$$

Equations (2.37) and (2.38) have to be solved for two parameters chosen from the plant model uncertainties or/and controller gains. Analytic solutions are difficult to obtain except for low-order plants; therefore, the use of a computer-aided solution procedure is advised. Gridding frequency $\omega \in [0, \sigma \tan \theta]$ and following a point-by-point numerical or symbolic procedure can achieve the solution. It is possible to make frequency ω go to infinity in the computations, but this will introduce fake solutions, which can easily be rejected. The fake solution computation also introduces unnecessary computational burden and is, therefore, not recommended.

The RRB for boundary ∂_1 degenerates into the single and real equation:

$$p(-\sigma, \boldsymbol{q}, \boldsymbol{k}) = \sum_{j=0}^{n+m} a_j(\boldsymbol{q}, \boldsymbol{k})(-1)^j \sigma^j = 0. \qquad (2.39)$$

The algebraic equation (2.39) has to be solved for two chosen parameters for each value of frequency ω. The solution at each frequency point will be a curve or a line. There is only the CRB for boundary ∂_2, which is given by

$$p(re^{j\theta}, q, k) = 0$$

with $\sigma/\cos\theta \leq r \leq R$ where gridding over r is required. The CRB for boundary ∂_2 can also be obtained from

$$p\left(\omega \frac{e^{j\theta}}{\sin\theta}, q, k\right) = p(\omega\cot\theta + j\omega, q, k) = 0$$

with $\sigma\tan\theta \leq \omega \leq R\sin\theta$ where gridding over ω can be used. This results in the real and imaginary component equations:

$$\mathrm{Re}\{p(\omega\cot\theta + j\omega, q, k)\} = 0, \tag{2.40}$$

$$\mathrm{Im}\{p(\omega\cot\theta + j\omega, q, k)\} = 0, \tag{2.41}$$

where $\sigma\tan\theta \leq \omega \leq R\sin\theta$.

The CRB for the boundary ∂_3 can be obtained from

$$p(Re^{j\theta}, q, k) = 0 \tag{2.42}$$

with R being fixed and θ varies within $\pi - \cos^{-1}\zeta_{\text{des}} \leq \theta \leq \pi$. This results in the real and imaginary component equations:

$$\mathrm{Re}\{p(Re^{j\theta}, q, k)\} = 0, \tag{2.43}$$

$$\mathrm{Im}\{p(Re^{j\theta}, q, k)\} = 0, \tag{2.44}$$

which need to be solved simultaneously for two uncertain parameters and/or controller gains. The RRB for the boundary ∂_3 degenerates into the single real equation:

$$p(-R, q, k) = \sum_{j=0}^{n+m} a_j(q, k)(-1)^j R^j = 0. \tag{2.45}$$

The algebraic equation (2.45) has to be solved for two chosen parameters for each value of frequency ω. The solution at each frequency point will be a curve or line.

Example 2.5 (Computation of $\mathscr{D}$-stability region)
Consider the plant

$$G(s) = \frac{1}{s^2 + 7s + 10}, \tag{2.46}$$

with the controller

$$C(s) = k_1 + \frac{k_2}{s}, \tag{2.47}$$

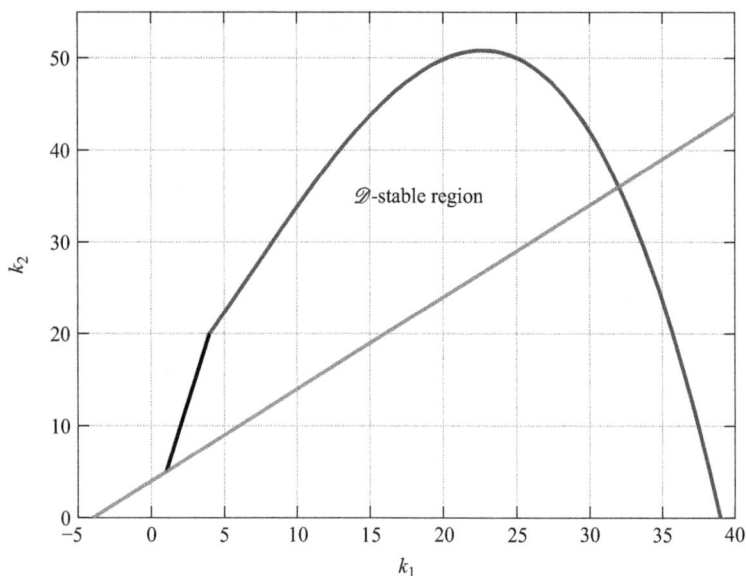

Figure 2.9   Parameter space solution for Example 2.5

where k_1 and k_2 are controller gains to be tuned, and $H(s) = 1$. The $\mathscr{D}$-stability boundary parameters are specified as $\sigma = 1$, $R = 7.5$ and $\theta = 60°$, i.e., $\zeta = \cos 60° = 1/2$. The $k_1 - k_2$ parameter space and the corresponding $\mathscr{D}$-stability bounding area in the complex plane corresponding to this example are shown in Figure 2.9. This parameter space was computed using the interactive parameter space file `RobustControlDesign.m` which is available in the website for the book.

2.4   Frequency domain control basics

Consider the feedback control system depicted in Figure 2.1. The loop gain of that block diagram is the product of all transfer functions in the loop given by

$$L(s) \triangleq C(s)G(s)H(s). \tag{2.48}$$

The closed-loop transfer function from reference input $R(s)$ to error $E(s)$ is

$$S(s) \triangleq \frac{E(s)}{R(s)} = \frac{1}{1 + L(s)} = \frac{1}{1 + C(s)G(s)H(s)}, \tag{2.49}$$

which is the sensitivity transfer function of the closed-loop system. One of the most important goals of feedback control is sensitivity minimisation to reduce error while tracking a reference input. The transfer function relations between the inputs $R(s)$, $D(s)$ and $N(s)$ and the output $Y(s)$ are given by

$$Y(s) = \underbrace{\frac{L(s)}{1+L(s)}}_{T(s)} \frac{1}{H(s)} R(s) + \underbrace{\frac{1}{1+L(s)}}_{S(s)} D(s) - \underbrace{\frac{L(s)}{1+L(s)}}_{T(s)} N(s), \qquad (2.50)$$

where $T(s) \triangleq L(s)/(1+L(s))$ is the complementary sensitivity function with $S(s) + T(s) = 1$. Conventional control shapes the open-loop gain $L(s)$ by using the controller $C(s)$ to indirectly minimise the closed-loop sensitivity and make the closed-loop complementary sensitivity equal to unity. This results in good command following and disturbance rejection but creates a conflict for sensor noise rejection, which requires the complementary sensitivity $T(s)$ to become zero. This conflict can be solved in conventional control by performing sensitivity minimisation (command following and disturbance rejection) at low frequencies and complementary sensitivity minimisation (sensor noise rejection) at high frequencies. The algebraic constraint $S(s) + T(s) = 1$ makes it impossible for both minimisations to be achieved simultaneously in the same frequency range. Therefore, one can assume that there will not be significant sensor noise within the bandwidth of the feedback controlled system. In fact, this assumption is, in general, reasonable. For the remaining cases when this assumption is not satisfied, the control system designer should go for higher quality sensors with a higher bandwidth of operation. The required loop gain Bode diagrams are illustrated in Figure 2.10.

2.4.1 Mapping phase margin bounds to parameter space

In Figure 2.10, ω_{gc} is the gain crossover frequency, i.e., the frequency at which the loop gain is unity or zero decibels while ω_{pc} is the phase crossover frequency, i.e., the frequency at which the phase is $-180°$. The phase margin PM and the gain margin GM are also shown in Figure 2.10. The polar or Nyquist plot representation in Figure 2.11 is also commonly used to illustrate the meaning of gain and phase crossover frequencies and phase and gain margins. This section concentrates on the phase margin, which necessitates the use of the gain crossover frequency in the computations.

The gain crossover frequency ω_{gc} is obtained by solving

$$|L(j\omega_{gc}, \boldsymbol{q}, \boldsymbol{k})| = 1, \qquad (2.51)$$

and the phase margin can be expressed as the solution of

$$\angle L(j\omega_{gc}, \boldsymbol{q}, \boldsymbol{k}) = e^{j(\mathrm{PM}-\pi)} = -\cos(\mathrm{PM}) - j\sin(\mathrm{PM}). \qquad (2.52)$$

Complex equation (2.52) results in the real and imaginary component equations:

$$\mathrm{Re}\{L(j\omega, \boldsymbol{q}, \boldsymbol{k})\} = -\cos(\mathrm{PM}), \qquad (2.53)$$

$$\mathrm{Im}\{L(j\omega, \boldsymbol{q}, \boldsymbol{k})\} = -\sin(\mathrm{PM}), \qquad (2.54)$$

where the gain crossover frequency ω_{gc} has been replaced by ω to let the gain crossover frequency be a free variable in the design procedure. The solution procedure is to grid

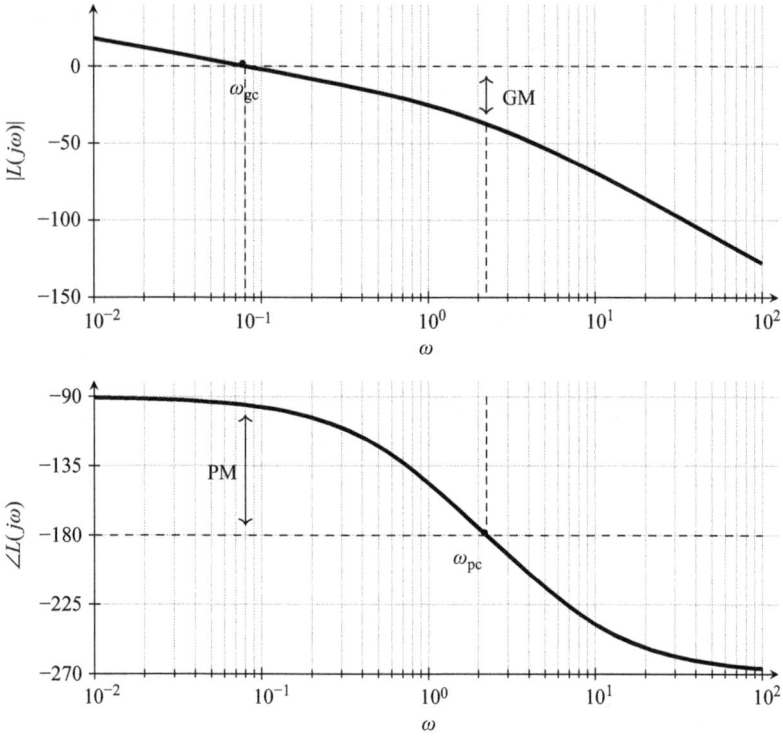

Figure 2.10   Loop gain requirements

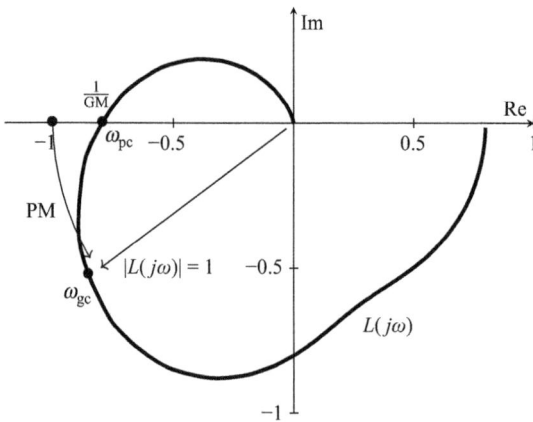

Figure 2.11   Nyquist plot representation

frequency ω, solve the two equations (2.53) and (2.54) for each value of ω for two unknown parameters in $q \in \mathbb{Q}$ and $k \in \mathbb{K}$. The procedure is very similar to the CRB solution procedure. The resulting points corresponding to each ω are plotted in the chosen parameter space. The desired phase margin PM is chosen by the designer before the solution procedure.

Example 2.6

Consider the feedback control system, shown in Figure 2.1, where the following PI controller,

$$C(s) = k_p + \frac{k_i}{s} = \frac{k_p s + k_i}{s}, \tag{2.55}$$

is used to regulate the plant,

$$G(s) = \frac{K}{Ts + 1} e^{-\tau_d s}, \tag{2.56}$$

which is of the first order and has a dead-time τ_d seconds, with a perfect sensor dynamics, i.e., $H(s) = 1$. Indeed, the plant (2.56) was chosen with a dead-time on purpose, to show that the phase margin bound mapping to parameter space can handle time delays in the system very easily as al computations are carried out in the frequency domain. The gain margin bound and sensitivity transfer function bound mapping to parameter space methods that are presented afterwards also have that property. The design equations (2.53) and (2.54) become

$$\text{Re}\{L(j\omega, q, k)\} = \text{Re}\left\{\frac{K(k_p j\omega + k_i)}{j\omega(Tj\omega + 1)} e^{-\tau_d j\omega}\right\} = -\cos(\text{PM}), \tag{2.57}$$

$$\text{Im}\{L(j\omega, q, k)\} = \text{Im}\left\{\frac{K(k_p j\omega + k_i)}{j\omega(Tj\omega + 1)} e^{-\tau_d j\omega}\right\} = -\sin(\text{PM}). \tag{2.58}$$

For this simple example, analytical solutions can be obtained by hand or by using symbolic manipulation as

$$k_p = -\frac{\cos(\text{PM} + \tau_d \omega) - T\omega \sin(\text{PM} + \tau_d \omega)}{K}, \tag{2.59}$$

$$k_i = \omega \frac{\sin(\text{PM} + \tau_d \omega) + T\omega \cos(\text{PM} + \tau_d \omega)}{K}. \tag{2.60}$$

Note that it is not always easy to obtain an analytical solution in which case it is easier to program the solution in MATLAB. In the MATLAB solution, the frequency ω is gridded and the two equations (2.57) and (2.58) are solved for the PI controller gains k_p and k_i for each frequency. The loci of solutions obtained are plotted in the $k_p - k_i$ parameter space to end the solution. The plant parameters

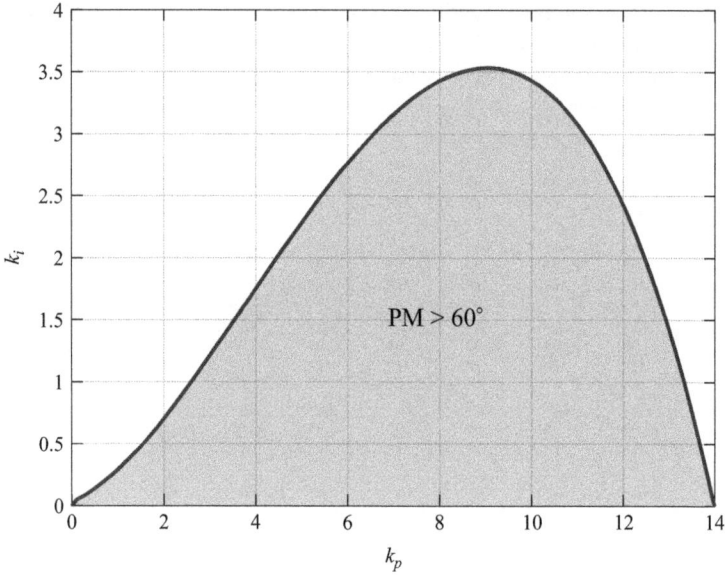

Figure 2.12   Phase margin bound of Example 2.6

K, T and τ_d are assumed to be known. The frequencies for which both k_p and k_i are positive values can be found using (2.59) and (2.60). The boundary and region in the k_p and k_i parameter space where phase margin PM > 60° is shown in Figure 2.12. The plant parameters $K = 12$, T $= 10$ s and $\tau_d = 0.2$ s were used in the computations for Figure 2.12. Symbolic manipulation was used to obtain (2.59) and (2.60), and to solve them for k_p and k_i at each frequency.

2.4.2   Mapping gain margin bounds to parameter space

The phase crossover frequency ω_{pc} is obtained by solving

$$\angle L(j\omega_{pc}, \boldsymbol{q}, \boldsymbol{k}) = -180°, \tag{2.61}$$

and the gain margin can be expressed as the solution of

$$L(j\omega_{pc}, \boldsymbol{q}, \boldsymbol{k}) = \frac{1}{\text{GM}}\angle{-180°} = -\frac{1}{\text{GM}}. \tag{2.62}$$

The complex equation (2.62) results in the real and imaginary component equations

$$\text{Re}\{L(j\omega, \boldsymbol{q}, \boldsymbol{k})\} = -\frac{1}{\text{GM}}, \tag{2.63}$$

$$\text{Im}\{L(j\omega, \boldsymbol{q}, \boldsymbol{k})\} = 0. \tag{2.64}$$

where the phase crossover frequency ω_{pc} has been replaced by ω to let the phase crossover frequency be a free variable in the design procedure. The solution procedure is to grid frequency ω, solve the two equations (2.63) and (2.64) for each value of ω for two unknown parameters in $q \in \mathbb{Q}$ and $k \in \mathbb{K}$. The procedure is very similar to the CRB solution procedure. The resulting points corresponding to each ω are plotted in the chosen parameter space. The desired gain margin GM is chosen by the designer before the solution procedure.

Example 2.7

Consider the following plant (2.56) and PI controller (2.55) of Example 2.6 with $H(s) = 1$. The plant parameters are chosen as $K = 12$, $T = 10$ s and $\tau_d = 0.2$ s as in Example 2.6. The design equations (2.63) and (2.64) become

$$\text{Re}\{L(j\omega, q, k)\} = \text{Re}\left\{\frac{K(k_p j\omega + k_i)}{j\omega(Tj\omega + 1)}e^{-\tau_d j\omega}\right\} = -\frac{1}{\text{GM}}, \tag{2.65}$$

$$\text{Im}\{L(j\omega, q, k)\} = \text{Im}\left\{\frac{K(k_p j\omega + k_i)}{j\omega(Tj\omega + 1)}e^{-\tau_d j\omega}\right\} = 0. \tag{2.66}$$

For this simple example, analytical solutions can be obtained by hand or by using symbolic manipulation as

$$k_p = -\frac{1}{\text{GM}}\frac{\cos(\tau_d\omega) - T\omega\sin(\tau_d\omega)}{K}, \tag{2.67}$$

$$k_i = \frac{1}{\text{GM}}\frac{\sin(\tau_d\omega) + T\omega\cos(\tau_d\omega)}{K}. \tag{2.68}$$

The frequencies for which both k_p and k_i are positive values can be found by using (2.67) and (2.68), and used in the MATLAB file for parameter space determination. The boundary and region in the $k_p - k_i$ parameter space where the gain margin GM > 2 is shown in Figure 2.13. Symbolic manipulation was used to obtain (2.67) and (2.68), and to solve them for k_p and k_i at each frequency.

It is possible to do multi-objective design by superimposing phase and gain bounds in the same parameter space. If the superimposition results in no intersection of solution regions, this means that no solution satisfies both criteria at the same time. Figure 2.14 shows the superimposed solution for PM > 60° and GM > 2 for the system in Examples 2.6 and 2.7.

2.4.3 Mapping sensitivity bounds to parameter space

Conventional control techniques concentrates on minimisation of sensitivity $S(s)$ as

$$\| W_S(s)S(s) \|_\infty \triangleq \sup_{\omega \in \mathbb{R}} |W_S(j\omega)S(j\omega)| < 1 \tag{2.69}$$

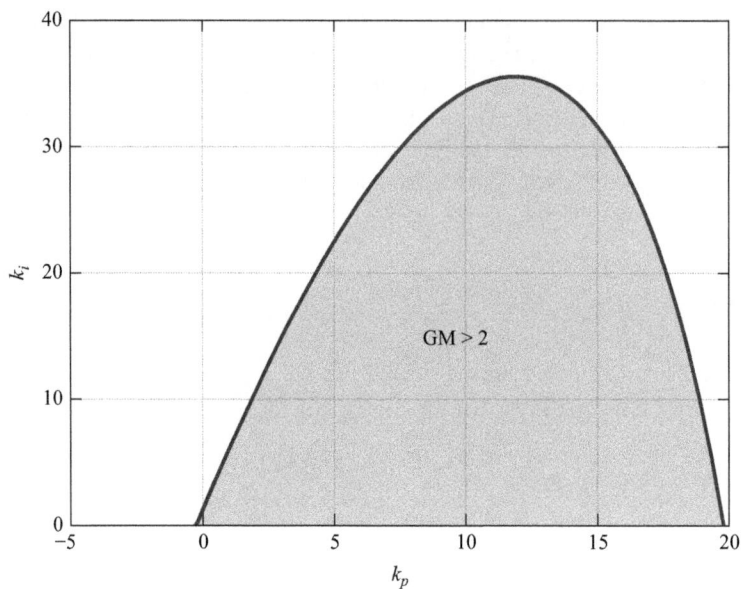

Figure 2.13   Gain margin bound of Example 2.7

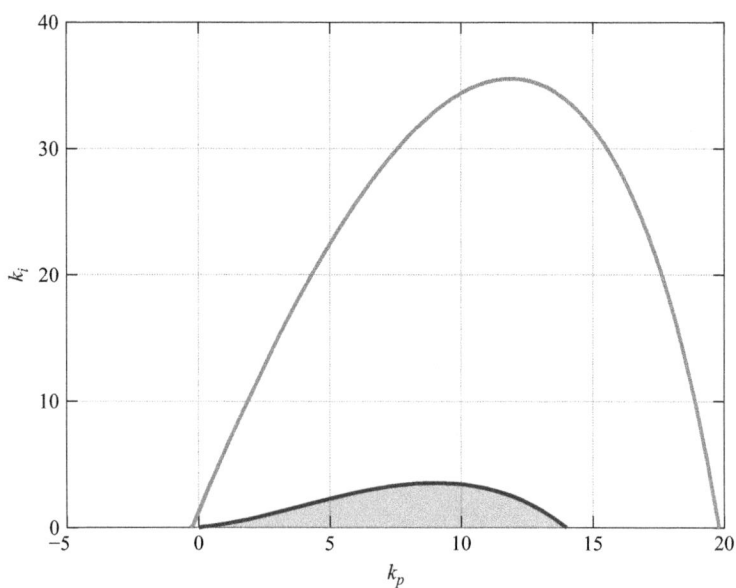

Figure 2.14   Intersection of phase margin and gain margin bounds of Example 2.8

to achieve desired command following and disturbance rejection within the bandwidth of the control system. Equation (2.69) is called nominal performance. The infinity norm is used in (2.69). The weighting function $W_S(s)$ is used to shape the sensitivity transfer function $S(s)$ and is usually chosen as the inverse of a desired sensitivity function $S_{\text{desired}}(s)$ as

$$W_S(s) = S_{\text{desired}}^{-1}(s). \tag{2.70}$$

A simple first-order rational transfer function that can be used for the sensitivity weight $W_S(s)$ is given in terms of its inverse as

$$W_S^{-1}(s) = h_s \frac{s + \omega_s l_s}{s + \omega_s h_s}, \tag{2.71}$$

where ω_s is the bandwidth of the desired sensitivity frequency response and l_s and h_s are its low-frequency (below bandwidth) and high-frequency (above bandwidth) gain, respectively. Even though real rational transfer function weights are required in $\mathcal{H}_\infty$ control, we will see that this restriction will not be applicable to the sensitivity bound mapping to parameter space design presented in this section. The sensitivity minimisation problem will be viewed in its equivalent form as

$$|W_S(j\omega)S(j\omega, \boldsymbol{q}, \boldsymbol{k})| < 1, \quad \forall \omega \in [0, \infty), \tag{2.72}$$

where the uncertain parameters and controller gains to be tuned have been shown as $\boldsymbol{q} \in \mathbb{Q}$ and $\boldsymbol{k} \in \mathbb{K}$. The equality version of (2.72), i.e.,

$$|W_S(j\omega)S(j\omega, \boldsymbol{q}, \boldsymbol{k})| = 1 \quad \text{for some } \omega, \tag{2.73}$$

will be called the *point condition* for nominal performance. The point condition provided in (2.73) will be mapped to the parameter space of two chosen parameters in $\boldsymbol{q} \in \mathbb{Q}$ and $\boldsymbol{k} \in \mathbb{K}$ for nominal performance.

Robustness of stability in the presence of a multiplicative model uncertainty $\Delta_m(s)$ such that the plant becomes

$$G(s) = G_n(s)(1 + W_T(s)\Delta_m(s)), \tag{2.74}$$

where $G_n(s)$ is the nominal plant model without uncertainty and the weight $W_T(s)$ is used to shape the multiplicative unstructured uncertainty $\Delta_m(s)$ such that $\|\Delta_m(s)\|_\infty < 1$. The stability robustness is then achieved by

$$\| W_T(s)T(s) \|_\infty \overset{\Delta}{=} \sup_{\omega \in \mathbb{R}} |W_T(j\omega)T(j\omega)| < 1 \tag{2.75}$$

where $T(s)$ is the complementary sensitivity function and the weight $W_T(s)$ is chosen as the inverse of the maximum high-frequency model uncertainty in the system. A simple first-order rational transfer function that can be used for the complementary sensitivity weight $W_T(s)$ is

$$W_T(s) = h_T \frac{s + \omega_T l_T}{s + \omega_T h_T}, \tag{2.76}$$

where ω_T is the bandwidth of the desired complementary sensitivity frequency response and l_T and h_T are its low-frequency (below bandwidth) and high-frequency (above bandwidth) gain, respectively. It should again be noted that unlike $\mathcal{H}_\infty$ control, the design method presented in this section does not require real rational transfer functions for weights. The weights can be chosen as a vector of magnitudes as different frequencies. Equation (2.75) is called the *robust stability* condition. The equality version of (2.75) will result in

$$|W_T(j\omega)T(j\omega,\boldsymbol{q},\boldsymbol{k})| = 1, \quad \text{for some } \omega, \tag{2.77}$$

which is the point condition for robust stability.

Nominal performance and robust stability are traditionally combined to obtain the *robust performance* condition given by

$$\| \, |W_S(s)S(s)| + |W_T(s)T(s)| \, \|_\infty < 1 \tag{2.78}$$

in the frequency domain approach to robust control. The corresponding equality with

$$|W_S(j\omega)S(j\omega,\boldsymbol{q},\boldsymbol{k})| + |W_T(j\omega)T(j\omega,\boldsymbol{q},\boldsymbol{k})| = 1, \quad \text{for some } \omega, \tag{2.79}$$

is called the point condition for robust performance which can also be expressed as

$$\left| W_S(j\omega)\frac{1}{1 + L(j\omega,\boldsymbol{q},\boldsymbol{k})} \right| + \left| W_T(j\omega)\frac{L(j\omega,\boldsymbol{q},\boldsymbol{k})}{1 + L(j\omega,\boldsymbol{q},\boldsymbol{k})} \right| = 1, \tag{2.80}$$

or equivalently,

$$|W_S(j\omega)| + |W_T(j\omega)L(j\omega,\boldsymbol{q},\boldsymbol{k})| = |1 + L(j\omega,\boldsymbol{q},\boldsymbol{k})|, \tag{2.81}$$

for some ω in terms of the loop gain $L(j\omega,\boldsymbol{q},\boldsymbol{k})$. Equation (2.81) is another form of the point condition for robust performance, also called mixed sensitivity. While a symbolic solution procedure for solving (2.81) is possible, a numerical solution procedure is preferred in this section as it is very easy to formulate, and it has lower computational burden as compared to symbolic manipulation, especially for higher order plant models. Therefore, (2.81) is solved frequency by frequency for two chosen parameters in $\boldsymbol{q} \in \mathbb{Q}$ and $\boldsymbol{k} \in \mathbb{K}$. In contrast to the CRB computation for Hurwitz and $\mathcal{D}$-stability, solution at each frequency results in a curve or line in parameter space. The combination of all these curves/lines results in the parameter space solution region.

The robust performance point condition (2.81) is illustrated graphically in Figure 2.15 where

$$L(j\omega,\boldsymbol{q},\boldsymbol{k}) = |L(j\omega,\boldsymbol{q},\boldsymbol{k})|\angle\theta_L, \tag{2.82}$$

and θ_L is the phase angle of the loop gain $L(j\omega,\boldsymbol{q},\boldsymbol{k})$.

Application of the cosine rule to the shaded triangle in Figure 2.15 and simplification results in the quadratic equation in $|L(j\omega,\boldsymbol{q},\boldsymbol{k})|$ given by

$$(1 - |W_T(j\omega)|^2)|L(j\omega,\boldsymbol{q},\boldsymbol{k})|^2 + 2(\cos\theta_L - |W_S(j\omega)|$$

$$\times |W_T(j\omega)|)|L(j\omega,\boldsymbol{q},\boldsymbol{k})| + (1 - |W_T(j\omega)|^2) = 0. \tag{2.83}$$

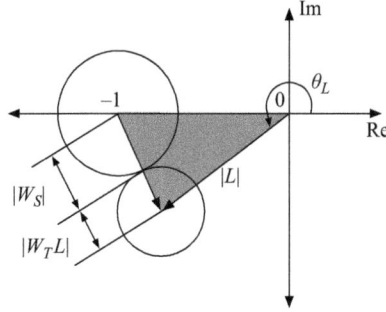

Figure 2.15   Robust performance point condition

The solution for $|L(j\omega, q, k)|$ based on the quadratic formula is given by

$$|L(j\omega, q, k)| = \frac{-\cos\theta_L + |W_S(j\omega)||W_T(j\omega)| \pm \sqrt{\Delta(\omega)}}{1 - |W_T(j\omega)|^2}, \qquad (2.84)$$

where

$$\Delta(\omega) = \cos^2\theta_L + |W_S(j\omega)|^2 + |W_T(j\omega)|^2$$
$$- 2|W_S(j\omega)||W_T(j\omega)|\cos\theta_L - 1. \qquad (2.85)$$

A numerical solution approach is used at each frequency ω by gridding and sweeping the angle θ_L from 0 to 2π radians and solving (2.84) and (2.85) for $|L|$ at each value of θ_L for which a solution exists. Then, all possible values of $L(j\omega) = |L(j\omega)|e^{j\theta_L}$ at the chosen frequency ω are obtained. Each value of $L(j\omega, q, k)$ satisfies

$$L(j\omega, q, k) = C(j\omega, k)G(j\omega, q)H(j\omega). \qquad (2.86)$$

The designer should grid frequency ω and grid θ_L for each value of ω and prepare point-by-point solution. This solution procedure is outlined below.

Algorithm 2.1

1. For $\omega_{min} \leq \omega \leq \omega_{max}$
   a. For $0 \leq \theta_L < 2\pi$
      i.   Solve (2.84) and (2.85) for $|L(j\omega)|$
      ii.   Discard non-real, non-positive $|L(j\omega)|$ solutions and keep the remaining
      iii. Calculate $L(j\omega) = |L(j\omega)|e^{j\theta_L}$
      iv. Substitute $L(j\omega)$ in (2.86) and solve for $C(j\omega, k)$ or $G(j\omega, q)$ or both
      v.   Solve for chosen two parameters in $q \in \mathbb{Q}$ and $k \in \mathbb{K}$
      vi. Plot solution points in parameter space
   b. End θ_L loop

2.   End ω loop
3.   Check all closed regions in the parameter space for solution regions that
satisfy $|W_S(j\omega)S(j\omega)| + |W_T(j\omega)T(j\omega)| < 1$, for all ω, and also check
$|W_S(j\omega)S(j\omega)| + |W_T(j\omega)T(j\omega)| = 1$, for some ω at the boundary of the
closed region

The design for nominal performance $W_S(j\omega)S(j\omega)$ for all $0 \le \omega < \infty$ is obtained
by setting $W_T(j\omega) = 0$ for all $0 \le \omega < \infty$ in (2.84) and (2.85), which results in the
single equation:

$$|L(j\omega, \boldsymbol{q}, \boldsymbol{k})| = -\cos\theta_L \pm \sqrt{\cos^2\theta_L + |W_S(j\omega)|^2 - 1}. \tag{2.87}$$

The design for robust stability $|W_T(j\omega)T(j\omega)| < 1$ for all $0 \le \omega < \infty$ is obtained
by setting $W_S(j\omega) = 0$ for all $0 \le \omega < \infty$ in (2.84) and (2.85), which results in the
single equation:

$$|L(j\omega, \boldsymbol{q}, \boldsymbol{k})| = \frac{-\cos\theta_L \pm \sqrt{\cos^2\theta_L + |W_T(j\omega)|^2 - 1}}{1 - |W_T(j\omega)|^2}. \tag{2.88}$$

The interactive m-file `RobustControlDesign.m` provided in the website for this
book has implemented $\mathscr{D}$-stability, phase margin bound and sensitivity bound mapping to parameter space objectives. This m-file can also be used for multi-objective
design by superimposing solutions for single objectives in parameter space.

2.5   Case study: automated path following

In this part of the chapter, an automated path following problem, described in
[15,16], is used as a numerical example to demonstrate parameter space controller
design for a family of plants represented by an uncertainty rectangle. The single-
track model, illustrated in Figure 2.16, is employed to model the steering dynamics.
Together with dynamics of the reference path and an actuator with integrating characteristics, the vehicle dynamics is characterised by a fifth-order state–space model:

$$\begin{bmatrix} \dot{\beta} \\ \dot{r} \\ \Delta\dot{\psi} \\ \dot{y} \\ \dot{\delta}_f \end{bmatrix} = \begin{bmatrix} a_{11} & a_{12} & 0 & 0 & b_{11} \\ a_{21} & a_{22} & 0 & 0 & b_{21} \\ 0 & 1 & 0 & 0 & 0 \\ v & l_s & V & 0 & 0 \\ 0 & -k_r & 0 & 0 & 0 \end{bmatrix} \begin{bmatrix} \beta \\ r \\ \Delta\psi \\ y \\ \delta_f \end{bmatrix} + \begin{bmatrix} 0 & 0 \\ 0 & 0 \\ 0 & -V \\ 0 & 0 \\ 1 & 0 \end{bmatrix} \begin{bmatrix} u_f \\ \rho_{\text{ref}} \end{bmatrix}, \tag{2.89}$$

where β, r, V, $\Delta\psi$, l_s, δ_f and y are vehicle side slip angle, vehicle yaw rate,
vehicle velocity, yaw angle relative to the desired path's tangent, preview distance,

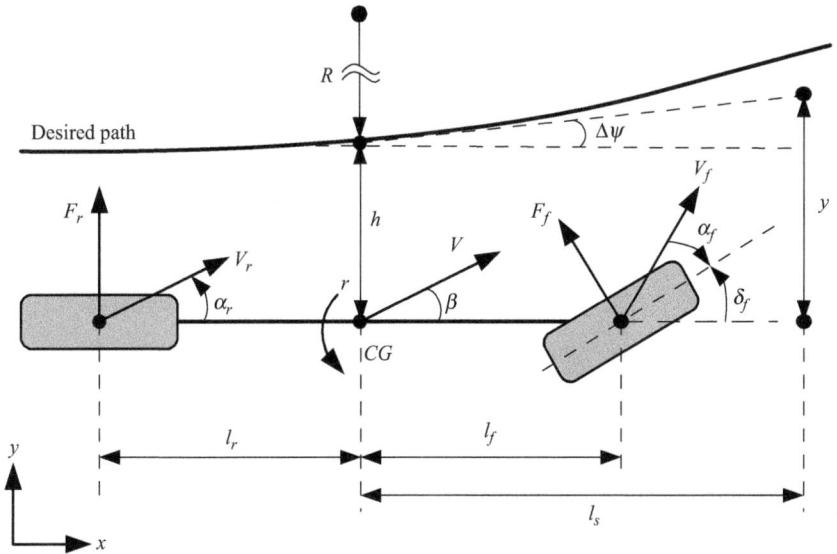

Figure 2.16 Illustration of vehicle path following model

steering angle and lateral deviation from the desired path at the preview distance, respectively. The non-zero entries of the matrices, provided in (2.89), are:

$$a_{11} = -\frac{c_r + c_f}{\tilde{m}V},$$

$$a_{12} = -1 + \frac{c_r l_r - c_f l_f}{\tilde{m}V^2},$$

$$a_{21} = \frac{c_r l_r - c_f l_f}{\tilde{J}},$$

$$a_{22} = -\frac{c_r l_r^2 + c_f l_f^2}{\tilde{J}V^2},$$

$$b_{11} = \frac{c_f}{\tilde{m}V},$$

$$b_{21} = \frac{c_f l_f}{\tilde{J}},$$

where $\tilde{m} \triangleq m/\mu$ is the virtual mass, $\tilde{J} \triangleq J/\mu$ is the virtual moment of inertia, μ is the road friction coefficient, m is the vehicle mass, J is the moment of inertia, c_f and c_r are the cornering stiffnesses, l_f is the distance from the centre of gravity of the vehicle (CG) to the front axle and l_r is the distance from the CG to the rear axle. The numerical values are taken from the product catalogue of Mercedes Sprinter: $l_f = 2.725\,\text{m}, l_r = 1.6\,\text{m}, l_s = 4.125\,\text{m}, m = 8{,}600\,\text{kg}, c_f = 100{,}000/\text{Nrad},$

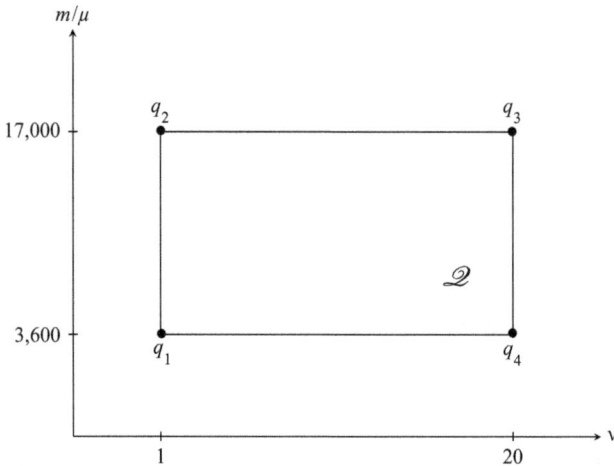

Figure 2.17   Operating domain of the vehicle

$c_f = 235,000/\text{Nrad}$ and $J = ml^2$ with $l^2 = 4.25\,\text{m}^2$. The vehicle mass, the vehicle velocity and the road friction coefficient are taken as uncertain parameters. The largest parametric variation occurs in the velocity V and virtual mass $\tilde{m}$ of the vehicle during the operation. The corresponding operating ranges of V and $\tilde{m}$ are illustrated in the (uncertainty) $\mathcal{Q}$-box (see Figure 2.17) where the corners of the $\mathcal{Q}$-box are labelled as q_i for all $i \in \{1, \ldots, 4\}$, and represent extreme points to be investigated in assessing robustness with respect to plant parameter variations.

A PID-type controller of the form

$$C(s) = \omega_c^3 \frac{k_{dd}s^3 + k_d s^2 + k_p s + k_i}{s(s^2 + 2\zeta\omega_c s + \omega_c^2)(s + \omega_c)} \tag{2.90}$$

is considered. The frequency ω_c is chosen as 100 rads while the damping ratio ζ is selected as 0.6. The proportional and integral gains are set to $k_p = 12.5$ and $k_i = 5$, respectively. A parameter space design procedure is performed to determine the derivative k_d and derivative squared k_{dd} gains in (2.90) for ensuring Hurwitz stability. The stability region computations are repeated for four different plants, each corresponding to one corner of the $\mathcal{Q}$-box of uncertainty in Figure 2.17. The Hurwitz stability regions, which are computed for each corner of $\mathcal{Q}$-box, are shown in Figure 2.18. The intersection of the Hurwitz stability regions is shaded in gray. The following controller:

$$C(s) = 100^3 \frac{k_{dd}s^3 + k_d s^2 + 12.5s + 5}{s(s^2 + 120s + 100^2)(s + 100)} \tag{2.91}$$

is chosen from the shaded region. This design corresponds to $k_d = 10$ and $k_{dd} = 0.5$. This controller will be robustly stable with respect to the chosen four plants within

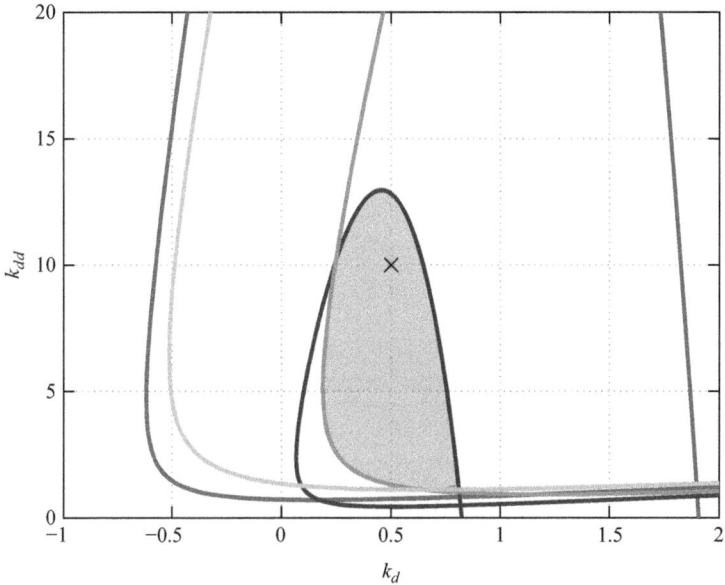

Figure 2.18   Parameter space region corresponding to Hurwitz stability

the parameter variations considered here. Although there is no theoretical guarantee, it is expected and hoped that Hurwitz stability will be maintained for all plants within the box of uncertainty in Figure 2.17 and not only the vertices.

The curvature, i.e., $\rho_{\mathrm{ref}} \triangleq 1/R$, of the guideline path is a measurable external disturbance. This external disturbance, which is the path to be followed, comprises of circular arcs with the transition to a new curvature corresponding to a step input in ρ_{ref}. Simulations are carried out for two different paths: Crows Landing and Richmond Field Station test tracks. When the vehicle travels on these tracks, the external disturbances illustrated in Figure 2.19 are applied. The simulation results are shown in Figures 2.20 and 2.21. A cruise controller was used in the simulations to maintain constant speed. The lateral displacement is always less than 0.2 m, which is viewed as satisfactory performance, in the simulations. Also, excessive yaw rate, which is another measure of the satisfactory performance, is not observed in the simulations.

2.6   Case study: Quanser QUBE™ Servo system

The Quanser QUBE™ servo is a commercially available educational system that consists of an electric motor connected to a rotational inertia load. It can also be converted to a rotary inverted pendulum. It consists of a brush d.c. motor with optical encoder position feedback, a pulse width modulation (PWM) amplifier and data

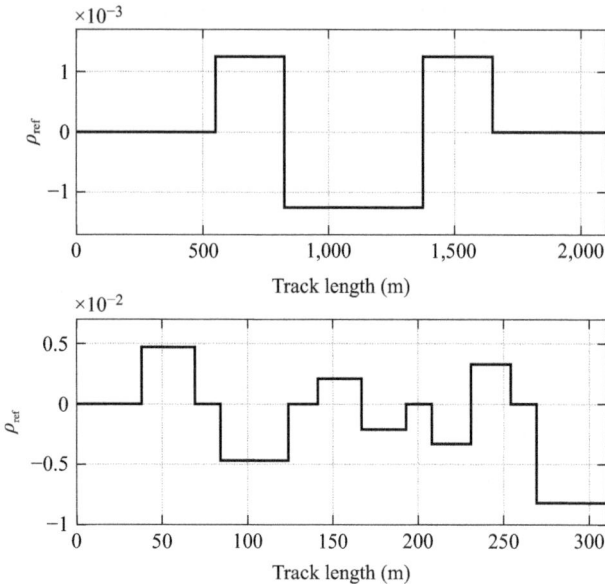

Figure 2.19   Radius of curvature, $\rho_{ref} = 1/R$, for two different test tracks

acquisition electronics which are all mounted within a compact aluminum housing. This educational plant comes with a Java program with fixed architecture controllers and has a rapid controller prototyping system called QUARC which allows students to be able to test their Simulink® controllers directly on the physical plant. It is, therefore, very easy to test the designed controllers in actual hardware. The modelling details and numerical values of the model parameters are given in [17].

This case study concentrates on the basic electric motor position and speed control for this plant. The plant from control input to motor speed is given by

$$G(s) = \frac{28}{0.1s + 1},\tag{2.92}$$

where the numerical values used were rounded to the nearest digit. A PI controller is used as the controller. A multi-objective parameter space design is presented here for this plant. The parameter space consists of the PI controller gains. The first objective is $\mathscr{D}$-stability. A $\mathscr{D}$-stability boundary with settling time/speed of response parameter $\sigma = 2$, $\alpha = 130°$ (i.e., corresponding to $\zeta = \cos 50° = 0.64$) and a bandwidth constraint of $R = 10$ rads were used. A frequency range between 1 rads and 30 rads with 100 frequency data points for the desired phase margin PM values in the range 70° to 90° with five levels, i.e., 70°, 75°, 80°, 85°, 90°, were used in the phase margin bound computations which form the second objective. The third objective is the achievement

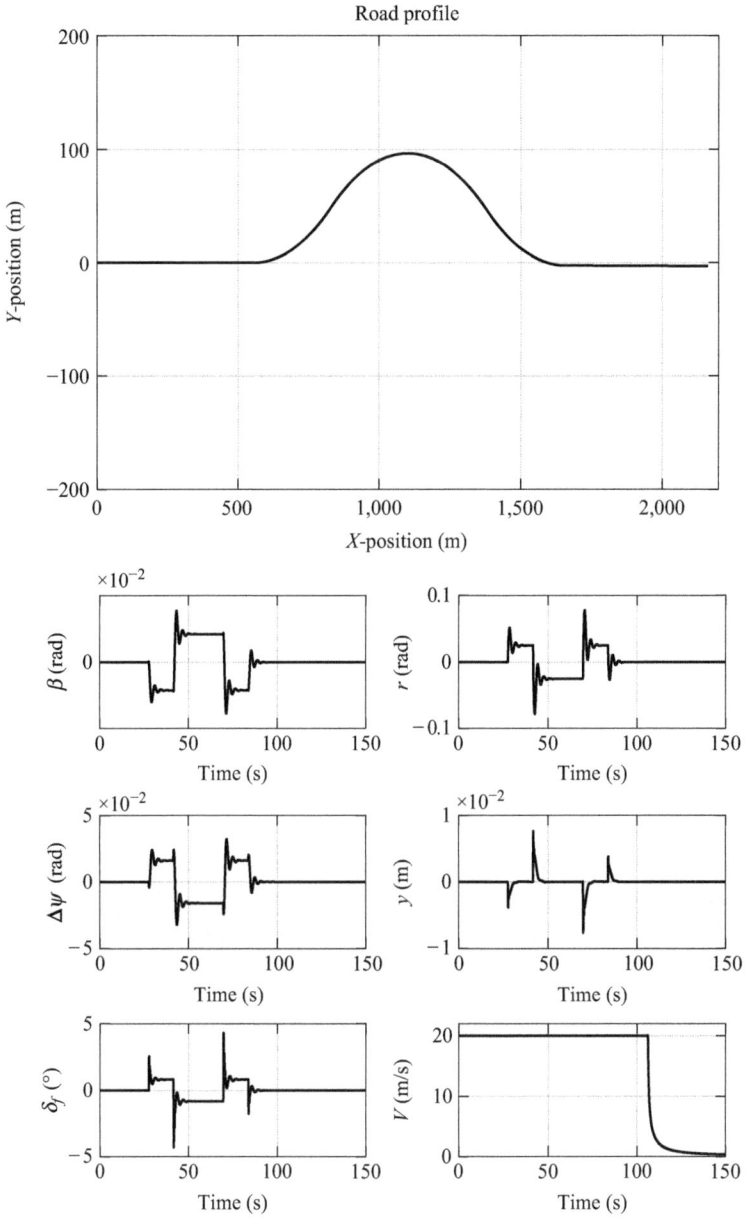

Figure 2.20   Simulation results for the first test track (Crows Landing)

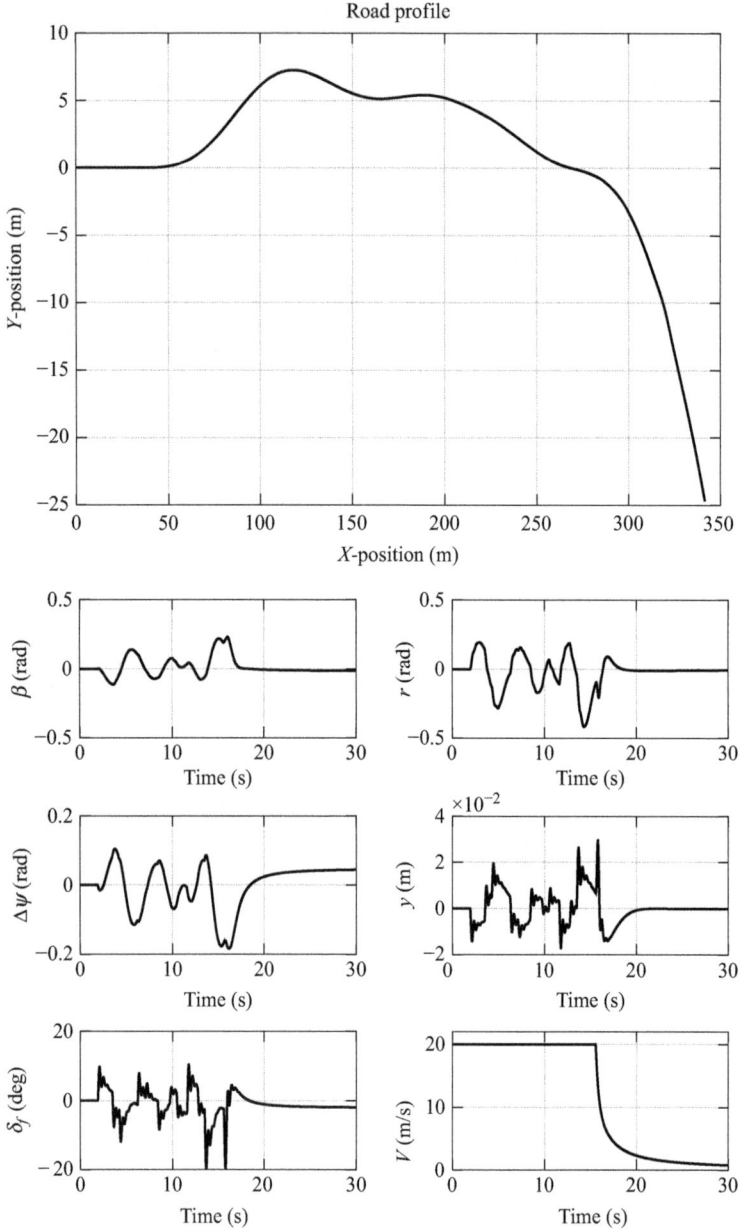

Figure 2.21   Simulation results for the second test track (Richmond Field Station)

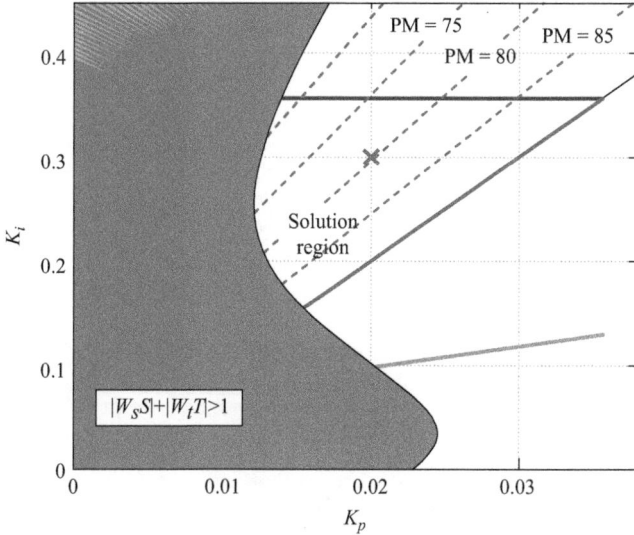

Figure 2.22   Multi-objective parameter space solution region

of the mixed sensitivity bound $|W_S(s)S(s)| + |W_T(s)T(s)| < 1$ with the sensitivity and complementary sensitivity weights

$$W_S^{-1}(s) = 2\frac{s + 6 \times 0.5}{s + 6 \times 2},\tag{2.93}$$

and

$$W_T(s) = 2\frac{s + 30 \times 0.2}{s + 30 \times 2},\tag{2.94}$$

respectively. The multi-objective solution region in the parameter space of PI controller gains is shown in Figure 2.22. The solution point chosen with controller gains $K_p = 0.02$ and $K_i = 0.3$ is shown with a cross sign in Figure 2.22. This point satisfies all objectives and has approximately 80° phase margin.

The closed-loop pole locations obtained are seen to be within the $\mathscr{D}$-stability boundary in Figure 2.23. The Bode plot in Figure 2.24 shows the achievement of about 80° of phase margin at a gain crossover frequency of about 7.5 rads. Figure 2.25 shows that the mixed sensitivity constraint $|W_S(s)S(s)| + |W_T(s)T(s)| < 1$ is met by this design. The step response of the system in Figure 2.26 shows that satisfactory speed regulation is obtained in response to a step command in desired speed of the motor.

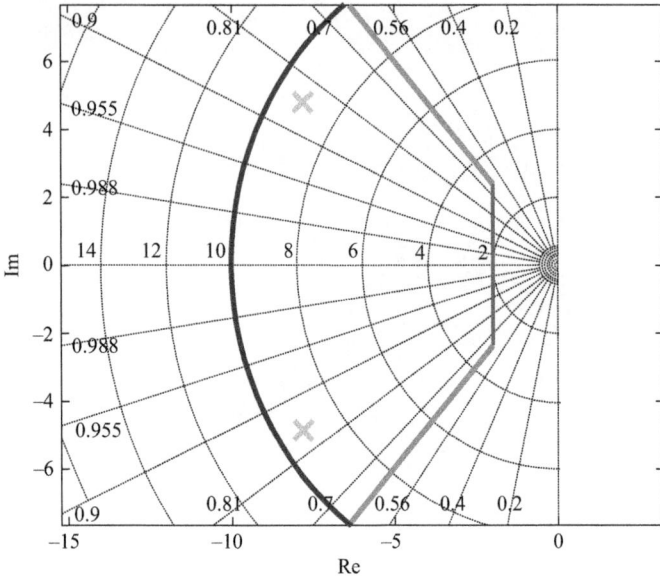

Figure 2.23   𝒟-stability region with closed-loop poles

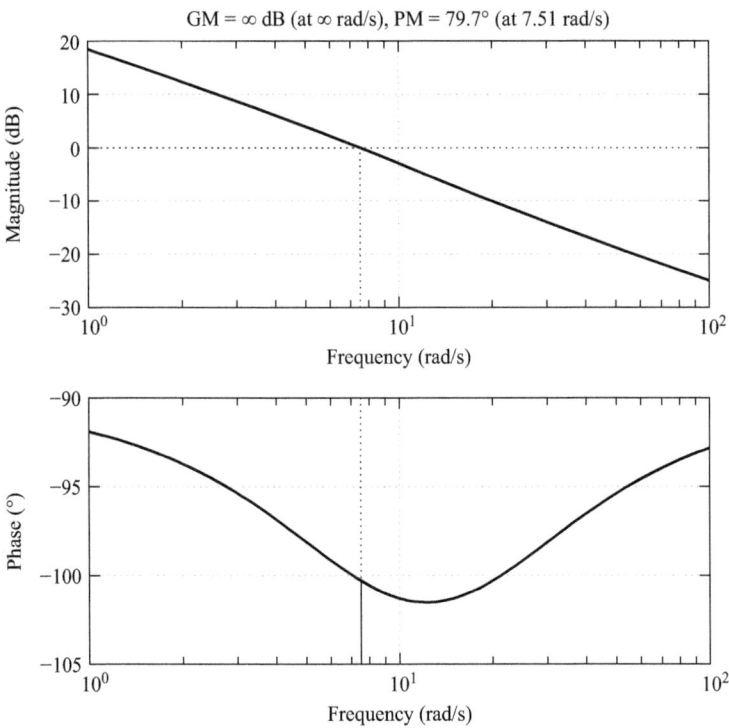

Figure 2.24   Bode plot showing phase margin achieved 79.7° at 7.51 rads

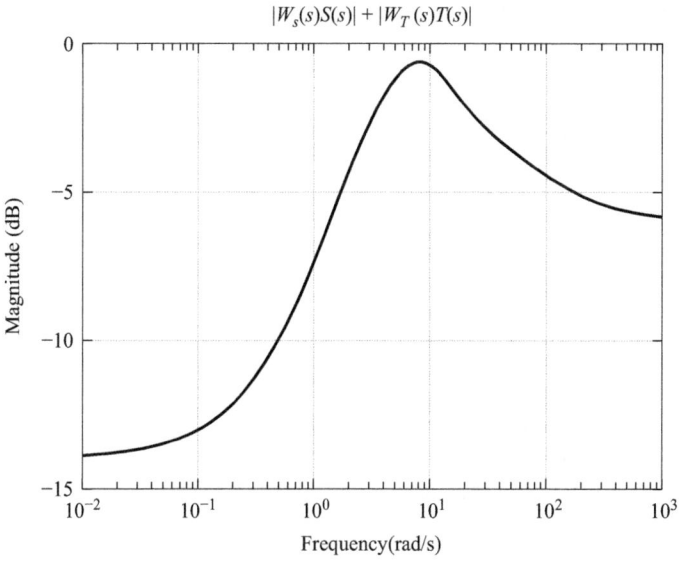

$$|W_s(s)S(s)| + |W_T(s)T(s)|$$

Figure 2.25   Mixed sensitivity plot

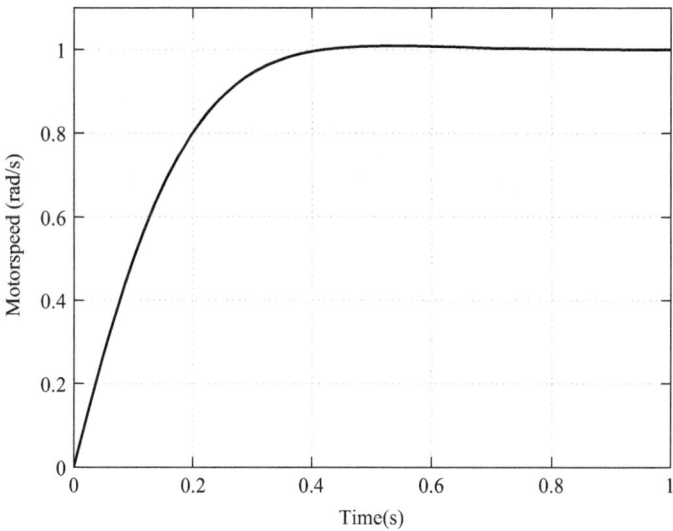

Figure 2.26   Step response of controlled system

2.7   Singular frequencies

Until this point in the chapter, the curves/lines in parameter space were obtained using a frequency sweep for Hurwitz and $\mathscr{D}$-stability using the boundary crossing theorem. In contrast, the condition at a single frequency was mapped into a whole curve or line in the mapping of frequency domain bounds into parameter space. A similar situation can arise for Hurwitz and $\mathscr{D}$-stability also when a curve or line is mapped into the parameter space at a singular frequency. This will occur when the two equations corresponding to real and imaginary components degenerate into a single equation. The trivial version of this was for the RRB and IRB when the two solution equations naturally became a single equation, meaning that $\omega = 0$ and $\omega = \infty$ are trivial cases of singular frequencies. In the non-trivial case for singular frequencies $\omega_s \in (0, \infty)$, the two CRB equations (2.7) and (2.8) for Hurwitz stability or CRB equations (2.37) and (2.38) for boundary ∂_1, CRB equations (2.40) and (2.41) for boundary ∂_2 and CRB equations (2.43) and (2.44) for boundary ∂_3 in the $\mathscr{D}$-stability computations should be checked for a dependence or equivalently a rank deficiency or determinant becoming zero in the solution procedure if a linear system of equations is obtained.

As a special case, singular frequencies occur in Hurwitz stable PID controllers. For fixed k_p values, the CRBs at the corresponding singular frequencies appear as straight lines in the $k_i - k_d$ parameter plane. These CRB lines together with RRB and IRB lines result in convex polygons in parameters space, which determine Hurwitz stable regions; see [12,18,19].

The characteristic equation of the PID-controlled system can be written as

$$p\left(s, k_p, k_i, k_d\right) = N(s)\left(k_p s + k_i + k_d s^2\right) + s D\left(s\right), \tag{2.95}$$

where $N(s)$ and $D(s)$ are the numerator and the denominator of the plant transfer function $G(s)$. Also, they can be written separately for their real and imaginary parts as $N(j\omega) = R_N + jI_N$ and $D\left(j\omega\right) = R_D + jI_D$.

Since $p\left(j\omega, k_p, k_i, k_d\right) = 0$ is a complex equation, it results in at most two independent equation as

$$R_p = \text{Re}(p(j\omega, k_p, k_i, k_d)) = 0, \tag{2.96}$$

$$I_p = \text{Im}(p(j\omega, k_p, k_i, k_d)) = 0. \tag{2.97}$$

For fixed k_p, (2.96) and (2.97) can be written in a matrix form as

$$\begin{bmatrix} R_p \\ I_p \end{bmatrix} = \underbrace{\begin{bmatrix} R_N & -R_N \omega^2 \\ I_N & -I_N \omega^2 \end{bmatrix}}_{M} \begin{bmatrix} k_i \\ k_d \end{bmatrix} + \begin{bmatrix} -k_p I_N \omega - I_D \omega \\ k_p R_N \omega + R_D \omega \end{bmatrix} = \begin{bmatrix} 0 \\ 0 \end{bmatrix}. \tag{2.98}$$

The matrix M in (2.98) is always singular and a solution for real singular frequencies (ω_s) can be obtained by solving the following equation:

$$R_N \left(k_p R_N \omega + R_D \omega\right) + I_N \left(k_p I_N \omega + I_D \omega\right) = 0. \tag{2.99}$$

For non-zero singular frequencies, one can obtain the straight CRB lines in the $k_i - k_d$ parameter plane by using the first row of (2.98) as follows:

$$k_i = k_d \omega_s^2 + \frac{k_p I_N(\omega_s)\omega_s + I_D(\omega_s)\omega_s}{R_N(\omega_s)}. \tag{2.100}$$

Example 2.8 (Singular frequencies in Hurwitz stable PID controller design)

Consider the plant

$$G(s) = \frac{N(s)}{D(s)} = \frac{s(s-0.5)}{s^3 + 3s^2 + 2s + 1} \tag{2.101}$$

with the PID controller

$$C(s) = k_p + \frac{k_i}{s} + k_d s^2, \tag{2.102}$$

where k_i and k_d are controller gains to be determined, k_p is a fixed value of 1.2 and $H(s) = 1$. For the given plant, the singular frequency can be calculated as 0.6014 by solving the equation given in (2.99). The corresponding CRB line is obtained using (2.100) as $k_i = 0.3617 k_d - 1.0383$.

Moreover, RRB for $\omega = 0$ and IRB for $\omega = \infty$ are obtained as $k_i = 2$, $k_d = -1$ lines by substituting corresponding frequency values into the characteristic equation. The parameter space solution region for $k_i - k_d$ pairs with RRB, IRB and CRB is shown in Figure 2.27. Also, closed-loop locations, which are Hurwitz stable, can be seen from Figure 2.28 for the selected test point controller pair.

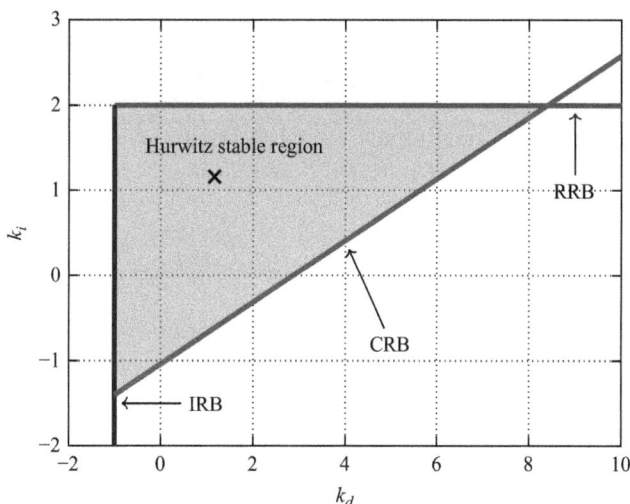

Figure 2.27 Parameter space solution for Example 2.8 (at the test point, i.e., $k_d = 1.1613$ and $k_i = 1.1587$)

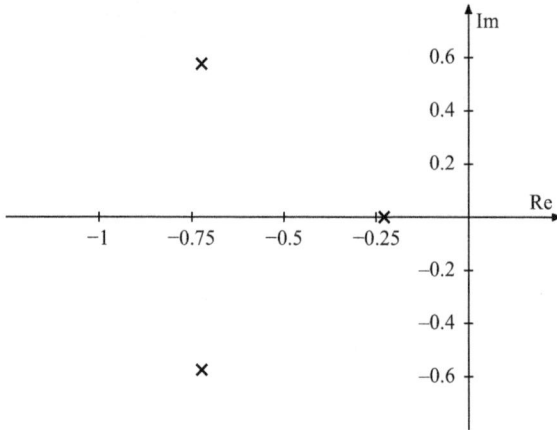

Figure 2.28   Closed-loop pole locations for the selected $k_i - k_d$ pair

References

[1] L. H. Keel and S. P. Bhattacharyya, "Robust, fragile, or optimal?" *IEEE Transactions on Automatic Control*, vol. 42, no. 8, pp. 1098–1105, Aug. 1997.

[2] K. Zhou, J. C. Doyle, and K. Glover, *Robust and Optimal Control*. Englewood Cliffs, NJ: Prentice Hall, 1996.

[3] G. Zames and J. G. Owen, "Duality theory for MIMO robust disturbance rejection," *IEEE Transactions on Automatic Control*, vol. 38, no. 5, pp. 743–752, May 1993.

[4] G. Zames, "Feedback and optimal sensitivity: Model reference transformations, multiplicative seminorms, and approximate inverses," *IEEE Transactions on Automatic Control*, vol. 26, no. 2, pp. 301–320, April 1981.

[5] J. C. Doyle, "Analysis of feedback systems with structured uncertainties," *IEE Proceedings D (Control Theory and Applications)*, vol. 129, no. 6, pp. 242–250, Nov. 1982.

[6] V. Kharitonov, "On a generalization of a stability criterion," *Seria fiziko-matematicheskaa, Izvestia Akademii nauk Kazakhskoi SSR*, vol. 1, pp. 53–57, 1978.

[7] A. C. Bartlett, C. V. Hollot, and H. Lin, "Root locations of an entire polytope of polynomials: It suffices to check the edges," *Mathematics of Control, Signals and Systems*, vol. 1, no. 1, pp. 61–71, Feb. 1988.

[8] W. Sienel and T. Bünte, "Design of gain scheduling controllers in parameter space," in *Proceedings of European Control Conference*, 1997.

[9] D. D. Šiljak, "Parameter space methods for robust control design: A guided tour," *IEEE Transactions on Automatic Control*, vol. 34, no. 7, pp. 674–688, July 1989.

[10] J. Ackermann, A. Bartlett, D. Kaesbauer, W. Sienel, and R. Steinhauser, *Robust Control: Systems with Uncertain Physical Parameters*, ser. Communications and Control Engineering. London, UK: Springer Verlag, 1993.

[11] J. Ackermann, "Parameter space design of robust control systems," *IEEE Transactions on Automatic Control*, vol. 25, no. 6, pp. 1058–1072, Dec. 1980.

[12] J. Ackermann, P. Blue, T. Bünte, L. Güvenç, D. Kaesbauer, M. Kordt, M. Mühler, and D. Odhental, *Robust Control: The Parameter Space Approach*. Springer Verlag: London, UK, 2002.

[13] R. A. Frazer and W. J. Duncan, "On the criteria for the stability of small motions," *Proceedings of The Royal Society A (Mathematical, Physical and Engineering Sciences)*, vol. 124, no. 795, pp. 642–654, July 1929.

[14] S. Gutman and E. I. Jury, "A general theory for matrix root-clustering in subregions of the complex plane," *IEEE Transactions on Automatic Control*, vol. 26, no. 4, pp. 853–863, Aug. 1981.

[15] B. Aksun-Güvenç, "Applied robust motion control," Ph.D. dissertation, Istanbul Technical University, Istanbul, Turkey, 2001.

[16] J. Ackermann, J. Guldner, W. Sienel, R. Steinhauser, and V. I. Utkin, "Linear and nonlinear controller design for robust automatic steering," *IEEE Transactions on Control Systems Technology*, vol. 3, no. 1, pp. 132–143, March 1995.

[17] (2017, Jan.). [Online]. Available: http://www.quanser.com/courseware/qubeservo_matlab/

[18] M.-T. Ho, A. Datta, and S. Bhattackaryya, "Design of P, PI and PID controllers for interval plants," in *Proceedings of American Control Conference*, 1998.

[19] M. T. Söylemez, N. Munro, and H. Baki, "Fast calculation of stabilizing PID controllers," *Automatica*, vol. 39, no. 1, pp. 121–126, Jan. 2003.

Chapter 3

Classical control

Conventional control methods, frequently used in the industry, including PID controllers, phase lead and lag compensators, and their variations are treated in this chapter. These are the first control methods that should be applied to a new problem, and if they solve the control design specifications satisfactorily, there is no need to try more advanced control methods or alternative control architectures like two-degrees-of-freedom control, for example. These classical techniques that work with Single-Input-Single-Output (SISO) transfer functions and root locus or frequency domain design methods are called conventional (or traditional) control systems here for lack of better terminology. We refrain from the use of classical control systems as techniques like state variable feedback and LQR that used to be known as modern control, and techniques like $\mathscr{H}_\infty$ control and μ-synthesis that used to be called new approaches. These have now become classical approaches themselves as decades have passed since their introduction to the controls literature. Conventional control techniques work best for SISO, Linear-Time-Invariant (LTI) systems where the required performance specifications are given in the time-domain and/or frequency-domain. Despite the presence of a vast number of advanced control techniques in the literature, conventional control methods are widely used in the control of mechatronic systems because they can easily be implemented as real-time systems and their costs are relatively cheaper as compared to more advanced techniques.

In this chapter, conventional control methods such as lead, lag, lead–lag compensators, PI, PD and PID controllers are reviewed first. This is followed by analytical solution techniques for their design based on desired phase margin, gain crossover frequency, and error constant combination. The use of existing numerical optimisation based PID and controller parameter tuners in MATLAB® and Simulink® is presented next. The chapter ends with the application of the multi-objective parameter space design method of Chapter 2 to PID controller design as a case study. All three of the proportional, integral and derivative gains are plotted in a three-dimensional parameter space in this case study.

3.1 Introduction to conventional control

There are two different approaches to reshaping the transient response of a closed-loop control system. One of them is the root-locus design approach, and the other one is the frequency response design approach. The root-locus design approach directly results in information on the transient response of the closed-loop control system as it displays the closed-loop pole locations as they vary with a chosen gain. In the frequency domain, the transient response is specified indirectly regarding the phase margin, gain margin, the resonant peak magnitude and the bandwidth, for example. These characteristics are related to the speed of response in the transient state and the static error constant in the steady state [1].

Trial-and-error procedures [2–8] are extensively used in designing conventional controllers such as lead, lag or lag–lead compensator in the frequency-domain. The main drawback of this heuristic frequency domain conventional controller design method is the trial-and-error phase. First, the designer selects a desired gain crossover frequency and phase margin. The designer, then, has to guess the phase lead (or phase lag) which will be needed at the new gain crossover frequency whose location is not known before the design process is completed as these compensators affect both phase and gain at the same time. Several trial-and-error cycles might be necessary before the desired phase margin is obtained. Experienced designers can usually guess the controller parameters, which will result in the desired level of performance of the closed-loop system after a few trials. However, inexperienced designers will be able to choose the correct parameters after many iterations. Indeed, the trial-and-error procedure is very heuristic and relies heavily on graphical frequency response plots.

The organisation of the rest of the chapter is as follows: Section 3.2 is on phase lead compensator design using the frequency domain analytical solution approach. Analytical frequency domain design of PD control is introduced as a special case of phase lead compensation. Section 3.3 is on the analytical solution of the phase lag compensator design in the frequency domain. PI control is treated as a special case of phase lead control. Analytical frequency domain design of phase lag-lead control is treated in Section 3.4 where PID control is treated as a special case. Numerical optimisation based MATLAB and Simulink tuning of PID control parameters is treated in Section 3.5. In Section 3.6, we present multi-objective parameter space design of a PID controller using a case study. Section 3.7 is dedicated to the case study of Quanser QUBE™ servo. The chapter ends with conclusions in Section 3.8.

3.2   Phase lead compensation

The phase lead compensator is a first-order filter whose phase angle Bode plot has a significant positive phase angle. It is used to add an extra phase lead (i.e., positive phase) to the controlled system. Also, the phase lead compensator usually increases the phase margin of the open-loop control system. The transient response of the system is improved, and small improvements in steady-state accuracy can also be achieved. Phase lead compensators will amplify high-frequency noise due to their gain characteristics.

3.2.1 Characteristics of phase lead compensators

The basic phase lead compensator, which contains a gain, pole and zero, has the following transfer function:

$$C_{\text{lead}}(s) = K_c \alpha \frac{Ts + 1}{\alpha Ts + 1} = K_c \frac{s + \frac{1}{T}}{s + \frac{1}{\alpha T}}, \tag{3.1}$$

where α is the attenuation factor, T is the compensator time constant, and K_c is the compensator gain. The compensator has one zero at $s = -1/T$ and one pole at $s = -1/\alpha T$. Since it has only one pole and one zero, they are both located on the real axis. The zero is always located to the right of the pole in the complex plane, i.e.,

$$-\frac{1}{\alpha T} < -\frac{1}{T},$$

since $0 < \alpha < 1$. The pole of the compensator migrates along the negative real axis from $s = -1/T$ to infinity as the factor α decreases. The polar (Nyquist) plot of the phase lead compensator is depicted in Figure 3.1 while its Bode plots (when $K_c = 1$, $\alpha = 0.1$ and $T = 10$) are illustrated in Figure 3.2. The phase lead compensator is used to add positive phase to a system to improve its phase margin. Due to the physical construction of the lead compensator, the minimum value of α is limited to 0.05; see, e.g., [1]. This choice of α provides approximately $\phi_m = 65°$ phase lead. In Figure 3.1, the angle between the positive real axis and the tangent line drawn from the origin to the semicircle gives the maximum phase lead angle ϕ_m for a given value of α.

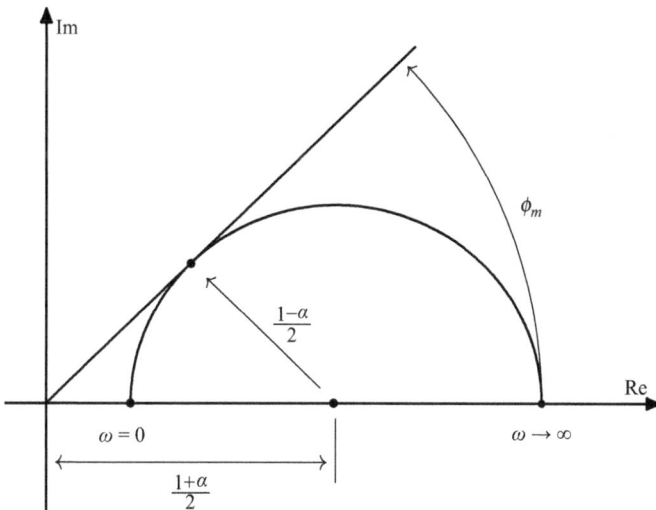

Figure 3.1   Polar plot of a phase lead compensator

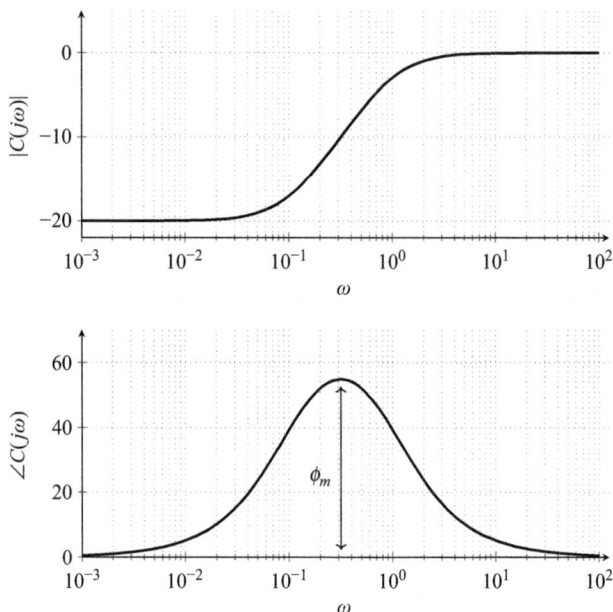

Figure 3.2   Bode plots of phase lead compensator ($K_c = 1$, $\alpha = 0.1$ and T $= 10$)

The frequency at the tangent is called ω_m. From Figure 3.1, the phase angle ϕ_m at $\omega = \omega_m$ is

$$\sin \phi_m = \frac{1 - \alpha}{1 + \alpha}. \tag{3.2}$$

3.2.2   Analytical phase lead compensator design in the frequency domain

To demonstrate the general use of the analytical solution procedure in the frequency domain, the procedure is applied to a time-delayed plant given by

$$G(s) = G_p(s)e^{-\tau_d s}. \tag{3.3}$$

The phase lead compensator is given by

$$C_{\text{lead}}(s) = K_c \alpha \underbrace{\frac{Ts + 1}{\alpha Ts + 1}}_{C_p(s)}, \tag{3.4}$$

where $C_p(s)$ is the normalised transfer function of the lead compensator, and it can be represented as

$$C_p(s) = \frac{Ts + 1}{\alpha Ts + 1}. \tag{3.5}$$

The equation that should be satisfied for obtaining the desired phase margin ϕ_m at the chosen gain crossover frequency ω_{gc} is

$$K_c\alpha C_p(j\omega_{gc})G_p(j\omega_{gc})H(j\omega_{gc})e^{-j\tau_d\omega_{gc}} = e^{j(-\pi+\phi_m)}. \tag{3.6}$$

The complex equation (3.6) results in the magnitude and angle equations:

$$|K_c\alpha C_p(j\omega_{gc})G_p(j\omega_{gc})H(j\omega_{gc})| = 1, \tag{3.7}$$

$$\theta \triangleq \angle C_p(j\omega_{gc}) = -\pi + \phi_m + \tau_d\omega_{gc} - \angle G_p(j\omega_{gc})H(j\omega_{gc}). \tag{3.8}$$

Equation (3.7) can be re-written as

$$|C_p(j\omega_{gc})| = \frac{1}{K_c\alpha|G_p(j\omega_{gc})H(j\omega_{gc})|}, \tag{3.9}$$

which is the gain condition that should be satisfied. Once the desired gain crossover frequency ω_{gc} is chosen and once $K_c\alpha$ is selected, the right-hand side of (3.9) is fixed and can be used to determine its left-hand side which is the magnitude of the normalised phase lead compensator at the gain crossover frequency. Once this magnitude is obtained, the normalised phase lead compensator becomes

$$C_p(j\omega_{gc}) = |C_p(j\omega_{gc})|e^{j\theta} \tag{3.10}$$

at the gain crossover frequency. Using (3.5), (3.9) and (3.10),

$$\frac{1+jT\omega_{gc}}{1+j\alpha T\omega_{gc}} = \frac{\cos\theta + j\sin\theta}{K_c\alpha|G_p(j\omega_{gc})H(j\omega_{gc})|} \tag{3.11}$$

is obtained. Using a common denominator on both sides of (3.11) and equating the resulting numerators results in

$$K_c\alpha|G_p(j\omega_{gc})H(j\omega_{gc})|(1+jT\omega_{gc}) = (1+j\alpha T\omega_{gc})(\cos\theta + j\sin\theta), \tag{3.12}$$

which can be re-expressed as

$$K_c\alpha|G_p(j\omega_{gc})H(j\omega_{gc})|(1+jT\omega_{gc}) = (\cos\theta - \alpha T\omega_{gc}\sin\theta)$$
$$+ j(\sin\theta + \alpha T\omega_{gc}\cos\theta). \tag{3.13}$$

The real and imaginary parts of the complex valued equation, provided in (3.13), are

$$\text{Real:} \quad \cos\theta - \alpha T\omega_{gc}\sin\theta = K_c\alpha|G_p(j\omega_{gc})H(j\omega_{gc})|, \tag{3.14}$$

$$\text{Imaginary:} \quad \sin\theta + \alpha T\omega_{gc}\cos\theta = K_c\alpha|G_p(j\omega_{gc})H(j\omega_{gc})|T\omega_{gc}. \tag{3.15}$$

By solving (3.14) and (3.15), the phase lead compensator's pole and zero locations are obtained as

$$-\frac{1}{\alpha T} = \frac{\sin\theta}{\cos\theta - K_c\alpha|G_p(j\omega_{gc})H(j\omega_{gc})|}\omega_{gc}, \tag{3.16}$$

$$-\frac{1}{T} = \frac{\sin\theta}{\frac{1}{K_c\alpha|G_p(j\omega_{gc})H(j\omega_{gc})|} - \cos\theta}\omega_{gc}. \tag{3.17}$$

When the desired phase margin ϕ_m and the desired gain crossover frequency ω_{gc} are selected, θ is known by (3.8) and the right-hand sides of (3.16) and (3.17) are known once $K_c\alpha$ is selected. Equations (3.16) and (3.17) are used to evaluate the pole and zero of the phase lead compensator, ending the design process.

The steps that should be taken for the analytical design of a phase lead compensator to satisfy the desired error constant, the desired gain crossover frequency and the desired phase margin are outlined next.

Procedure:

Step 1 Evaluate the compensator gain $K_c\alpha$ so that the desired error constant (e.g., K_p, K_v or K_a) is obtained.

Step 2 Generate the Bode diagram of the open-loop system $K_c\alpha G(s)H(s)$, and locate its gain crossover frequency $\tilde{\omega}_{gc}$. Choose the new gain crossover frequency ω_{gc} subject to $|K_c\alpha G(j\omega_{gc})H(j\omega_{gc})|$ having magnitude lower than 0 dB.

Step 3 Evaluate

$$\theta = -\pi + \phi_m + \tau_d\omega_{gc} + \angle G_p(j\omega_{gc})H(j\omega_{gc}),$$

and solve for the remaining phase lead compensator parameters by using

$$\frac{1}{\alpha T} = \frac{\sin\theta}{\cos\theta - K_c\alpha|G_p(j\omega_{gc})H(j\omega_{gc})|}\omega_{gc},$$

$$\frac{1}{T} = \frac{\sin\theta}{\frac{1}{K_c\alpha|G_p(j\omega_{gc})H(j\omega_{gc})|} - \cos\theta}\omega_{gc}.$$

Step 4 Draw the Bode plots of the compensated system

$$K_c\alpha\frac{Ts+1}{\alpha Ts+1}G(s)H(s).$$

Evaluate the phase and gain margins to make sure that the desired phase margin is obtained and that the resulting gain margin is satisfactory.

Note that several constraints must be satisfied for the analytical solution procedure presented above. These limitations are presented next.

Constraints for phase lead compensation:

- $0° < \theta < 90°$ is needed for being able to add positive phase lead since the upper limit is the theoretical maximum phase lead that can be provided by the phase lead compensator.
- $|G_p(j\omega_{gc})| < 1$ is needed to be able to change the gain crossover frequency as the phase lead compensator adds positive magnitude and hence moves the gain crossover frequency.
- $K_c\alpha|G_p(j\omega_{gc})H(j\omega_{gc})| < \cos\theta < \frac{1}{K_c\alpha|G_p(j\omega_{gc})H(j\omega_{gc})|}$ is needed for $1/T$ and $1/\alpha T$ to be positive.

Example 3.1

Consider the open-loop system, provided in [3], is given by

$$G(s) = \frac{1}{s(0.2s + 1)(0.45s + 1)} = \frac{1}{0.09s^3 + 0.65s^2 + s}. \tag{3.18}$$

The velocity error constant for the uncompensated plant is $1\,s^{-1}$. The gain crossover frequency of this plant is $0.91\,\text{rad/s}$ without any compensation. It is desired to design a compensator for the plant of (3.18) such that the static velocity error constant K_v is $1.8\,s^{-1}$, the phase margin is at least $70°$, and the gain crossover frequency is $4\,\text{rad/s}$. In other words, we are asking for a slight increase in the velocity error constant, a large phase margin and a large increase in gain crossover frequency which will also increase the controlled system's bandwidth. Using the Conventional Controller Design part of the COMES toolbox, the lead compensator that satisfies the desired design specifications in the frequency-domain will be designed.

In Figure 3.3, the general structure of the Graphical User Interface (GUI) based on MATLAB is demonstrated. On the right-hand side of Figure 3.3, the Bode magnitude and phase plots can be seen. In these plots, the black curves denote the uncompensated system while the gray curves indicate the compensated system. The type of controller, which will be designed, can be chosen from the *Controller Type* option in the menu bar. According to the selected controller type, the user entry sections seen on the left-hand side of Figure 3.3 are either activated or deactivated.

The resulting phase lead compensator is obtained as

$$C_{\text{lead}}(s) = \frac{2.598s + 1.8}{0.002486s + 1}. \tag{3.19}$$

It should be noted that as the desired velocity error constant is increased further in the design procedure, the phase lead compensator, described above, starts approaching a proportional-plus-derivative controller. The unit step response of the phase lead compensated system is displayed in Figure 3.4 along with the unit step response of a simple feedback controlled system. This simple feedback control consists of only adding a negative feedback loop around the plant which is the same as proportional control with a proportional gain of one. This will be called *basic feedback control* in all the examples in this chapter. The phase lead compensated system shows faster rise time and similar settling time without any overshoot for this design in comparison to basic feedback control, which is the simplest feedback controller that involves no tuning effort. The response in Figure 3.4 is typical of a slow mode due to one pole of the closed-loop system being close to the imaginary axis as shown in the closed-loop system pole-zero map of Figure 3.5. Note that the fast pole, introduced by the phase lead compensator, is not displayed in Figure 3.5.

The Bode plots of the phase lead compensated system are shown in Figure 3.6. It is verified that the desired phase margin and gain crossover frequency have been obtained. The resulting gain margin of $39.3\,\text{dB}$ at the phase crossover frequency of $51.3\,\text{rad/s}$ is also highly satisfactory.

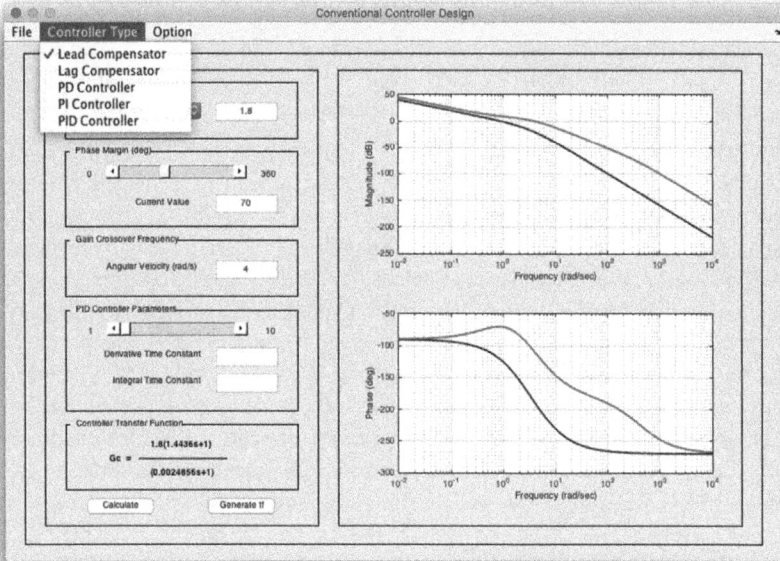

Figure 3.3   Phase lead compensator design in COMES toolbox

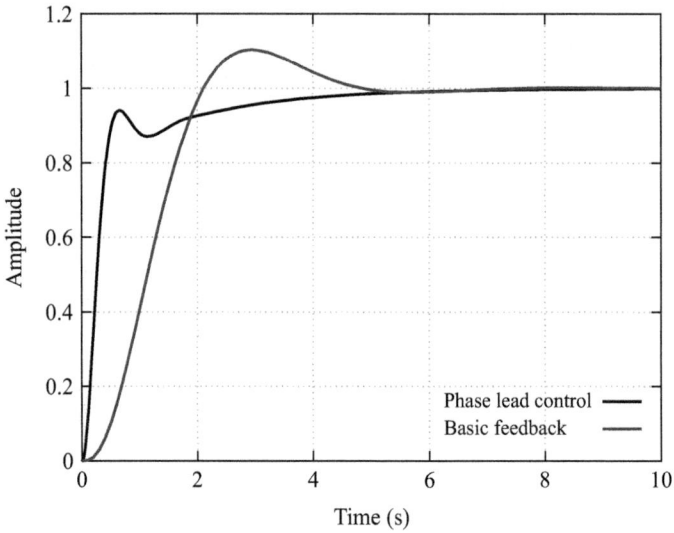

Figure 3.4   Unit step responses of phase lead and basic feedback controlled systems

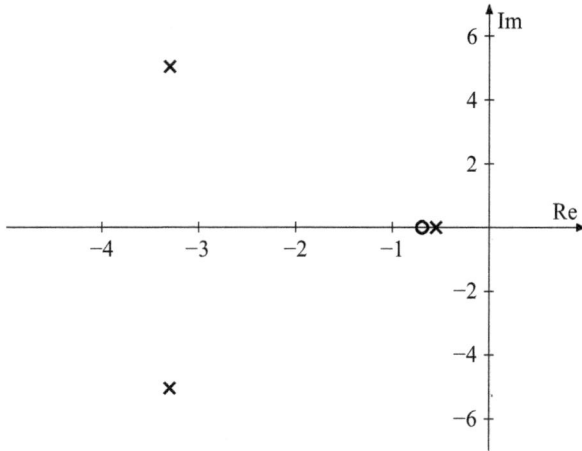

Figure 3.5   Pole-zero map of phase lead controlled system

GM = 39.3 dB (at 51.3 rad/s), PM=70° (at 4 rad/s)

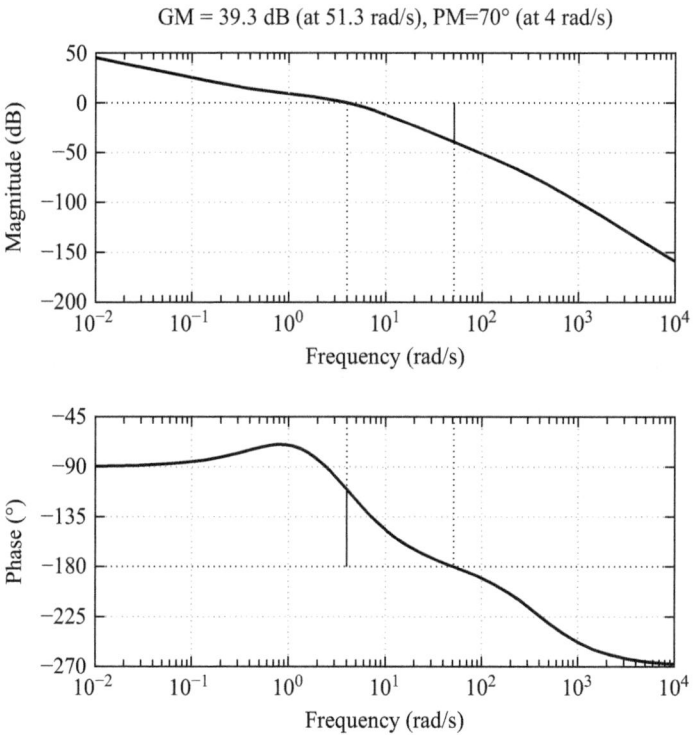

Figure 3.6   Bode plots of phase lead controlled system

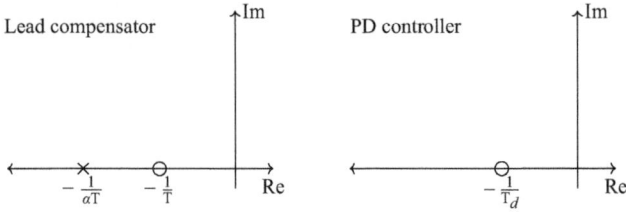

Figure 3.7   The pole-zero locations of phase lead compensator and PD controller

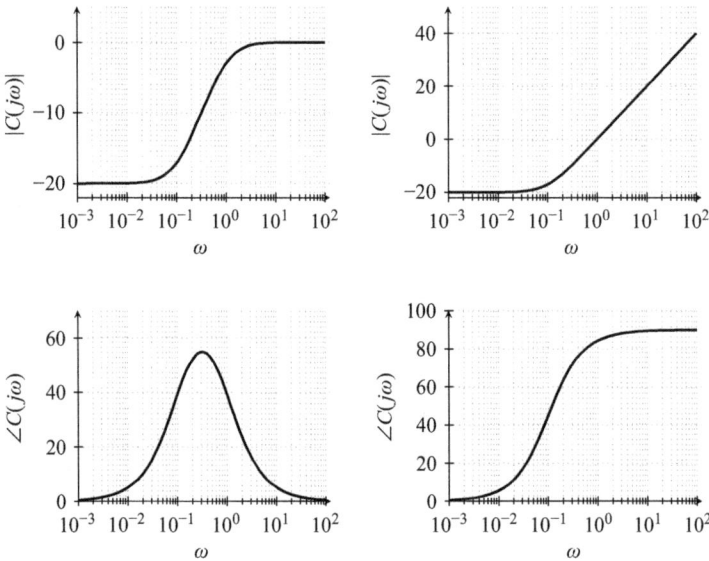

Figure 3.8   Bode plots of phase lead compensator (left) and PD controller (right)

3.2.3   PD control as a special case of phase lead compensation

The proportional-plus-derivative (PD) control is given by

$$C_{\text{PD}}(s) = k_p(T_d s + 1) = k_p T_d s + k_p = k_d s + k_p, \tag{3.20}$$

where k_p is the proportional gain, k_d is the derivative gain and T_d is the derivative time constant. A comparison of the controller's pole-zero patterns for PD and phase lead control systems, displayed in Figure 3.7, shows that PD control is a special case of phase lead compensator with its pole located at infinity, i.e., $\alpha \to 0$.

The Bode magnitude and phase angle plots of PD and phase lead controllers are drawn side-by-side in Figure 3.8. The Bode plots are identical at low frequency with differences at higher frequencies. The analytical design procedure is, therefore, very similar to that of the phase lead compensator and is presented next.

The two equations that must be satisfied for achieving the desired phase margin ϕ_m at the desired gain crossover frequency ω_{gc} are similar to (3.7) and (3.8) for a phase lead compensator and are given by

$$|C_{PD}(j\omega_{gc})G_p(j\omega_{gc})H(j\omega_{gc})| = |(k_p + jk_d\omega_{gc})G_p(j\omega_{gc})H(j\omega_{gc})| = 1, \quad (3.21)$$

$$\theta \triangleq \angle C_{PD}(j\omega_{gc}) = -\pi + \phi_m + \tau_d\omega_{gc} - \angle G_p(j\omega_{gc})H(j\omega_{gc}). \quad (3.22)$$

Equation (3.21) becomes

$$|C_{PD}(j\omega_{gc})| = |k_p + jk_d\omega_{gc}| = \frac{1}{|G_p(j\omega_{gc})H(j\omega_{gc})|}. \quad (3.23)$$

The frequency response of the controller is calculated as

$$C_{PD}(j\omega_{gc}) = |C_{PD}(j\omega_{gc})|e^{j\theta} = |k_p + jk_d\omega_{gc}|(\cos\theta + j\sin\theta). \quad (3.24)$$

Using (3.23) and (3.24), the PD controller can be expressed as

$$k_p + jk_d\omega_{gc} = \frac{\cos\theta + j\sin\theta}{|G_p(j\omega_{gc})H(j\omega_{gc})|}, \quad (3.25)$$

which can be manipulated into

$$(k_p + jk_d\omega_{gc})|G_p(j\omega_{gc})H(j\omega_{gc})| = \cos\theta + j\sin\theta. \quad (3.26)$$

The real and imaginary parts of the complex valued equation, provided in (3.26), can be obtained as follows:

$$\text{Real:} \qquad k_p|G_p(j\omega_{gc})H(j\omega_{gc})| = \cos\theta, \quad (3.27)$$

$$\text{Imaginary:} \qquad k_d|G_p(j\omega_{gc})H(j\omega_{gc})|\omega_{gc} = \sin\theta. \quad (3.28)$$

The solution of (3.27) and (3.28) results in the proportional and derivative gains of the PD controller:

$$k_p = \frac{\cos\theta}{|G_p(j\omega_{gc})H(j\omega_{gc})|}, \quad (3.29)$$

$$k_d = \frac{\sin\theta}{|G_p(j\omega_{gc})H(j\omega_{gc})|\omega_{gc}}, \quad (3.30)$$

and the derivative time constant is given by

$$T_d = \frac{k_d}{k_p} = \frac{\tan\theta}{\omega_{gc}}. \quad (3.31)$$

There are two parameters which are the gains k_p and k_d for a PD controller that can be tuned as compared to three tuneable parameters in a phase lead compensator. Consequently, the analytical design of a PD controller will aim for the achievement of the desired phase margin at the desired gain crossover frequency and will not involve a static error constant. The steps involved in this procedure is outlined next.

Procedure:

Step 1   Generate the Bode diagram of the open-loop system $G(s)H(s)$ and locate its gain crossover frequency $\tilde{\omega}_{gc}$. Choose the new gain crossover frequency ω_{gc} subject to $G(j\omega_{gc})H(j\omega_{gc})$ having magnitude lower than 0 dB.

Step 2   Evaluate

$$\theta = -\pi + \phi_m + \tau_d\omega_{gc} + \angle G_p(j\omega_{gc})H(j\omega_{gc}),$$

and solve for the PD controller gains by using

$$k_p = \frac{\cos\theta}{|G_p(j\omega_{gc})H(j\omega_{gc})|},$$

$$k_d = \frac{\sin\theta}{|G_p(j\omega_{gc})H(j\omega_{gc})|\omega_{gc}},$$

and

$$T_d = \frac{\tan\theta}{\omega_{gc}}.$$

Step 3   Draw the Bode plots of the compensated system

$$(k_d s + k_p)G(s)H(s).$$

Evaluate the phase margin and gain margin to make sure that the desired phase margin is obtained and that the resulting gain margin is satisfactory. Here, $0° < \theta < 90°$ should be satisfied for the analytical solution procedure to work as this is the angle range that the PD controller can have. If this is not satisfied, a different desired gain crossover frequency should be used.

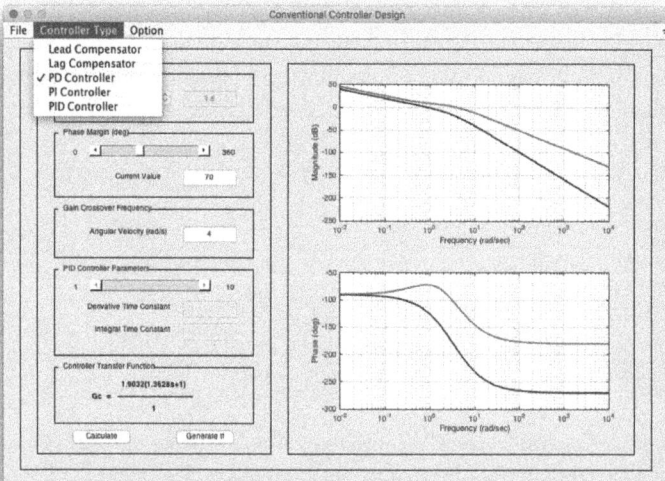

Figure 3.9   PD controller design in COMES toolbox

Example 3.2 (PD controller design)

Consider the same plant, as given in Example 3.1, which is

$$G(s) = \frac{1}{s(0.2s + 1)(0.45s + 1)}.$$

Let us use the same specification of a phase margin that is at least 70° at the desired gain crossover frequency of 4 rad/s. A static velocity error constant will not be specified as there are not enough control parameters that can be tuned to achieve that extra specification. By utilising the Conventional Controller Design part of the COMES toolbox, the PD controller that satisfies the desired design specifications is designed as illustrated in Figure 3.9.

The resulting PD controller is obtained as

$$C_{\text{PD}}(s) = k_d s + k_p = 2.592s + 1.903. \tag{3.32}$$

Note that the PD controller cannot be implemented as it is not causal. Instead of the mathematical differentiator in (3.32), a limited differentiator has to be used as

$$C_{\text{PD}}(s) = k_d s \frac{\omega_N}{s + \omega_N} + k_p = \frac{(k_d + k_p)s + k_p \omega_N}{s + \omega_N}, \tag{3.33}$$

where ω_N is the cut-off frequency of the limited differentiation.

The unit step response of the uncompensated and PD-controlled system are displayed in Figure 3.10. The PD-controlled system shows faster rise time and similar settling time at no overshoot for this design. The Bode plots with and without PD control are shown in Figure 3.11. It is verified that the desired phase margin and gain crossover frequency have been obtained. The resulting infinite gain margin that is achieved is also highly satisfactory.

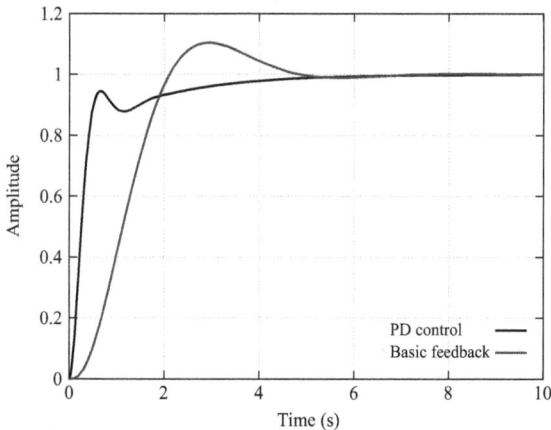

Figure 3.10   Unit step responses of PD and basic feedback controlled systems

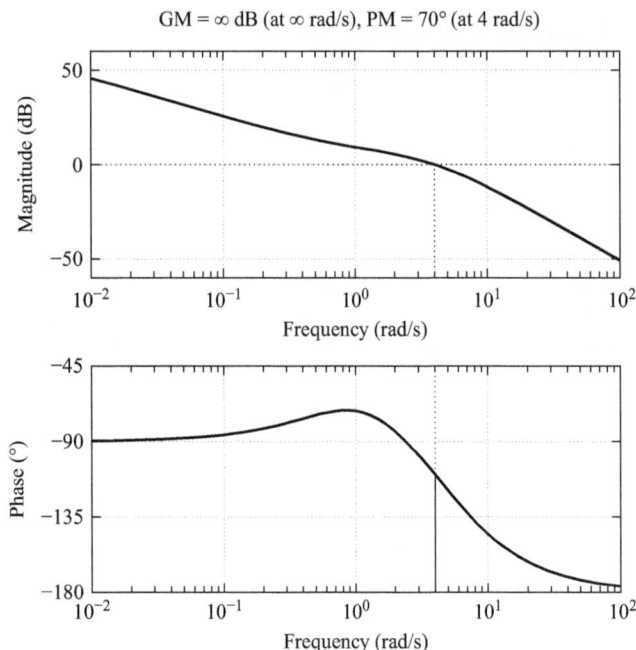

GM = ∞ dB (at ∞ rad/s), PM = 70° (at 4 rad/s)

Figure 3.11   Bode plots of PD-controlled system

3.3   Phase lag compensation

Phase lag compensators are employed to increase the steady-state accuracy without adversely affecting the overall dynamic response. Once a satisfactory dynamic response has been obtained, perhaps by the use of lead compensation, the designer may want to increase the value of the relevant error constant like the velocity error constant K_v, for example. For this reason, a pole close to the origin is introduced to approximate integration. Recall that PI control or I control alone reduces the steady-state error due to the high loop gain at low frequencies. A zero close to this pole is also introduced so that the pole-zero pair near the origin does not significantly interfere with the overall system dynamic response. This is the basis for the phase lag compensator design.

3.3.1   Characteristics of phase lag compensators

The transfer function of a phase lag compensator is given by

$$C_{\text{lag}}(s) = K_c\beta\frac{Ts+1}{\beta Ts+1} = K_c\frac{s+\frac{1}{T}}{s+\frac{1}{\beta T}}, \tag{3.34}$$

where $\beta > 1$ is an important factor for the lag compensator. In the complex plane, a phase lag compensator has a zero at $s = -1/T$ and a pole at $s = -1/\beta T$. The pole is

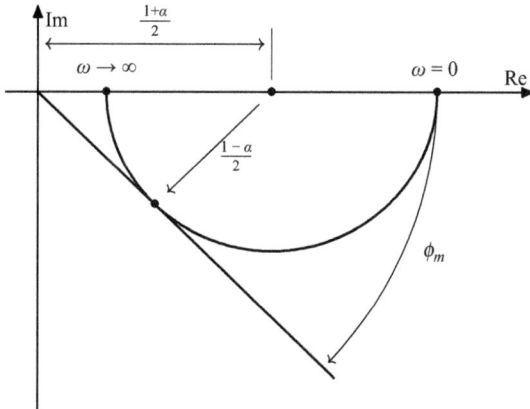

Figure 3.12 Polar plot of a phase lag compensator

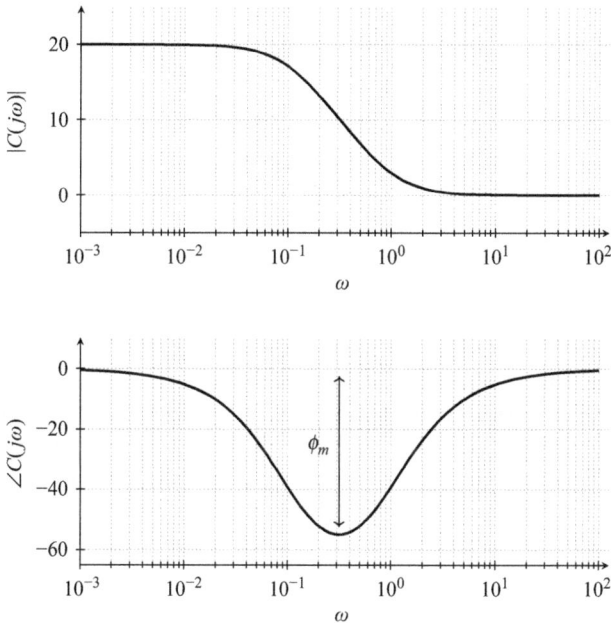

Figure 3.13 Bode plots of phase lag compensator ($K_c = 1$, $\alpha = 10$ and T = 1)

always located to the right of the zero due to the condition $\beta > 1$. Figure 3.12 shows the polar plot of a phase lag compensator. Figure 3.13 shows a Bode diagram of the phase lag compensator with $K_c = 1$ and $\beta = 10$. The phase lag compensator has a large negative phase angle which is not a desirable quality. So, conventional phase lag compensator design concentrates on low frequencies where that extra positive phase is not too large. The corner frequencies of the phase lag compensator in the Bode magnitude plot are at $\omega = 1/\tau$ and $\omega = 1/\beta\tau$. In Figure 3.13 the values of K_c and β are set equal to 1 and 10, respectively. The magnitude of the phase lag compensator becomes 20 dB at low frequencies and 0 dB at high frequencies. The phase lag compensator Bode diagram resembles that of a low-pass filter.

It is desired to have close to unity phase lag compensator gain at high frequencies ($K_c \approx 1$) so that the overall frequency response will not be affected at high frequencies. The additional negative phase that is introduced, i.e., the phase lag, is an undesirable property and the designer likes to work at frequencies where there is very small phase lag. This is in contrast to the lead compensator design procedure where the designer is most interested at the frequency where the maximum positive phase contribution (phase lead) occurs.

When we have a system whose open-loop transfer function has sufficient phase margin, and we desire to increase its steady-state accuracy, the use of a phase lag compensator at very low frequencies to increase the value of the relevant error constant is necessary. The phase lag compensator will be designed so as not to affect the rest of the system's dynamic response at higher frequencies. The resulting phase margin and gain crossover will not have appreciable changes. The conventional approach is first to design a phase lead compensator for the desired phase margin at the chosen gain crossover frequency. A phase lag compensator is then designed for the phase lead compensated plant so that the achieved phase margin and gain crossover frequency are preserved while the phase lag compensator improves the chosen static error constant, thus, improving accuracy. This is similar to adding an integrator to the system to improve steady-state error characteristics.

As compared to phase lead compensation, phase lag compensation usually results in lower bandwidth. Another motivation for the use of phase lag compensation is to achieve lower loop gain at higher frequencies where the attenuation of noise effects or noise immunity is desired. Lower loop gain at higher frequencies is also needed for improved stability robustness in the face of high-frequency modelling errors. The analytical frequency domain approach for designing a phase lag compensator is presented in the next subsection. This analytical solution is based on keeping a desired phase margin at the chosen gain crossover frequency while significantly improving low-frequency gain in the form of a larger static error constant. The idea is to keep close to the phase margin of the plant or previously phase lead compensated plant at a slightly lower bandwidth while increasing low-frequency gain.

3.3.2 Analytical phase lag compensator design in the frequency domain

The phase lag compensator can be expressed as

$$C_{\text{lag}}(s) = K_c \beta \underbrace{\frac{Ts + 1}{\beta Ts + 1}}_{C_n(s)}, \tag{3.35}$$

where the gain $K_c\beta$ of the phase lag compensator is separated from its unity d.c. gain part $C_n(s)$. The plant $G(s)$ is still given by (3.3) which consists of a dead-time τ_d and a real rational transfer function part $G_p(s)$. The equation that should be satisfied for obtaining the desired phase margin ϕ_m at the chosen gain crossover frequency ω_{gc} is

$$K_c\beta C_n(j\omega_{gc})G_p(j\omega_{gc})H(j\omega_{gc})e^{-j\tau_d\omega_{gc}} = e^{j(-\pi+\phi_m)}. \tag{3.36}$$

The complex equation (3.36) results in the magnitude and angle equations:

$$|K_c\beta C_n(j\omega_{gc})G_p(j\omega_{gc})H(j\omega_{gc})| = 1, \tag{3.37}$$

$$\theta \triangleq \angle C_n(j\omega_{gc}) = -\pi + \phi_m + \tau_d\omega_{gc} - \angle G_p(j\omega_{gc})H(j\omega_{gc}). \tag{3.38}$$

Equation (3.37) can be re-written as

$$|C_n(j\omega_{gc})| = \frac{1}{K_c\beta|G_p(j\omega_{gc})H(j\omega_{gc})|}, \tag{3.39}$$

which is the gain condition that should be satisfied. Once the desired gain crossover frequency ω_{gc} is chosen and once $K_c\beta$ is selected, the right-hand side of (3.39) is fixed and can be used to determine its left-hand side which is the magnitude of the normalised phase lag compensator at the gain crossover frequency. Once this magnitude is obtained, the phase lag compensator becomes

$$C_n(j\omega_{gc}) = |C_n(j\omega_{gc})|e^{j\theta} \tag{3.40}$$

at the gain crossover frequency. Using (3.35), (3.39) and (3.40),

$$\frac{1+jT\omega_{gc}}{1+j\beta T\omega_{gc}} = \frac{\cos\theta + j\sin\theta}{K_c\beta|G_p(j\omega_{gc})H(j\omega_{gc})|} \tag{3.41}$$

is obtained. Using a common denominator on both sides of (3.41) and equating the resulting numerators result in

$$K_c\beta|G_p(j\omega_{gc})H(j\omega_{gc})|(1+jT\omega_{gc}) = (1+j\beta T\omega_{gc})(\cos\theta + j\sin\theta), \tag{3.42}$$

which can be re-expressed as

$$K_c\beta|G_p(j\omega_{gc})H(j\omega_{gc})|(1+jT\omega_{gc}) = (\cos\theta - \beta T\omega_{gc}\sin\theta) + j(\sin\theta + \beta T\omega_{gc}\cos\theta). \tag{3.43}$$

The real and imaginary parts of the complex valued equation, seen in (3.43), are

$$\text{Real:} \quad \cos\theta - \beta T\omega_{gc}\sin\theta = K_c\beta|G_p(j\omega_{gc})H(j\omega_{gc})|, \tag{3.44}$$

$$\text{Imaginary:} \quad \sin\theta + \beta T\omega_{gc}\cos\theta = K_c\beta|G_p(j\omega_{gc})H(j\omega_{gc})|T\omega_{gc}. \tag{3.45}$$

By solving (3.44) and (3.45), the phase lag compensator's pole and zero locations are obtained as

$$-\frac{1}{\beta T} = -\frac{\sin\theta}{\cos\theta - K_c\beta|G_p(j\omega_{gc})H(j\omega_{gc})|}\omega_{gc}, \tag{3.46}$$

$$-\frac{1}{T} = -\frac{\sin\theta}{\dfrac{1}{K_c\beta|G_p(j\omega_{gc})H(j\omega_{gc})|} - \cos\theta}\omega_{gc}. \tag{3.47}$$

The necessary equations have been derived. When the desired phase margin ϕ_m and the selected gain crossover frequency ω_{gc} are selected, θ is known by (3.38), and the right-hand sides of (3.46) and (3.47) are known once $K_c\beta$ is selected. Equations (3.46) and (3.47) are used to evaluate the pole and zero of the phase lag compensator, ending the design process.

The steps that should be taken for the analytical design of a phase lag compensator to satisfy a desired error constant and to keep a desired gain crossover frequency and a desired phase margin are outlined next.

Procedure:

Step 1   Evaluate the compensator gain $K_c\beta$ so that the desired error constant (e.g., K_p, K_v or K_a) is satisfied.

Step 2   Generate the Bode diagram of the open-loop system $K_c\beta G(s)H(s)$, and locate its gain crossover frequency $\tilde{\omega}_{gc}$. Choose the new gain crossover frequency ω_{gc} to be slightly smaller than $\tilde{\omega}_{gc}$.

Step 3   Evaluate

$$\theta = -\pi + \phi_m + \tau_d\omega_{gc} + \angle G_p(j\omega_{gc})H(j\omega_{gc}),$$

and solve for the remaining phase lag compensator parameters by using

$$\frac{1}{\beta T} = \frac{\sin\theta}{\cos\theta - K_c\beta|G_p(j\omega_{gc})H(j\omega_{gc})|}\omega_{gc},$$

$$\frac{1}{T} = \frac{\sin\theta}{\frac{1}{K_c\beta|G_p(j\omega_{gc})H(j\omega_{gc})|} - \cos\theta}\omega_{gc}.$$

Step 4   Draw the Bode plots of the compensated system

$$K_c\beta\frac{Ts + 1}{\beta Ts + 1}G(s)H(s).$$

Evaluate the phase and gain margins to make sure that the desired phase margin is obtained and that the resulting gain margin is satisfactory.

Note that several constraints must be satisfied for the analytical solution procedure presented above. These limitations are presented next.

Constraints for phase lag compensation:

- $-90° < \theta < 0°$ is the largest possible range of angle values that the phase lag compensator can have.
- $\omega_{gc} < \tilde{\omega}_{gc}$ is needed since the phase lag compensator will reduce the bandwidth.
- $K_c\beta|G_p(j\omega_{gc})H(j\omega_{gc})| < \cos\theta < \frac{1}{K_c\alpha|G_p(j\omega_{gc})H(j\omega_{gc})|}$ is needed for $1/T$ and $1/\beta T$ to be positive.

Example 3.3 (Phase lag compensator design)

Consider the same plant, as given in Example 3.1, which is

$$G(s) = \frac{1}{s(0.2s + 1)(0.45s + 1)}.$$

It is desired to design a compensator for this plant such that the static velocity error constant is $K_v = 10\,\text{s}^{-1}$, the phase margin is at least $70°$ and the gain crossover frequency is $0.5\,\text{rad/s}$. The Conventional Controller Design part of the COMES toolbox is used as shown in Figure 3.14 to design a phase lag compensator. First, the static velocity error is entered, followed by the desired phase margin. Finally, the gain crossover frequency is entered, and the "*Calculation*" button is pressed to generate the controller. In Figure 3.14, the transfer function of the phase lag compensator calculated in the COMES toolbox is displayed in the bottom. The controller transfer function can be exported into the MATLAB workspace by pressing "*Generate tf*" button. The phase lag compensator obtained is

$$C_{\text{lag}}(s) = 10\frac{67.53s + 1}{1,312s + 1}. \tag{3.48}$$

Notice that $\beta = 53 > 1$.

The unit-step response and the Bode plots (with and without compensation) of the phase lag compensated system are displayed in Figures 3.15 and 3.16, respectively. The unit step response in Figure 3.15 shows a slower response with much smaller overshoot as compared to the uncompensated system. The Bode diagram of Figure 3.16 shows that the desired phase margin is obtained at the desired gain crossover frequency. There is a frequency range just below the gain crossover frequency where the phase is quite low due to the dip in the phase plot of the phase lag compensator phase Bode plot (see Figure 3.13). We avoid that part of the frequency response of a phase lag compensator and use the extra positive magnitude made available at very low frequencies.

Phase lag compensators usually reduce the closed-loop control system's bandwidth, resulting in a slower system. The speed of response could be increased by reducing β and/or by moving the phase lag compensator's pole closer to the origin. Sometimes, a phase lead compensator may need to be designed and added to achieve the desired phase margin in a preliminary design step. The phase lag compensator will then be designed for the phase lead compensated plant.

3.3.3   PI control as a special case of phase lag control

The proportional-plus-integral (PI) controller can be expressed as

$$C_{\text{PI}}(s) = k_p\left(1 + \frac{1}{T_i s}\right) = \frac{k_p\left(s + \frac{1}{T_i}\right)}{s} = \frac{k_p s + k_i}{s}, \tag{3.49}$$

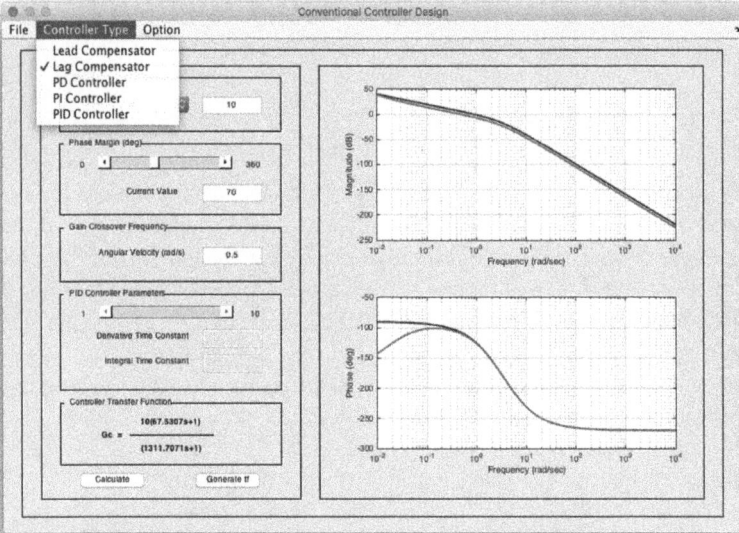

Figure 3.14   Phase lag compensator design using the conventional control part of the COMES toolbox

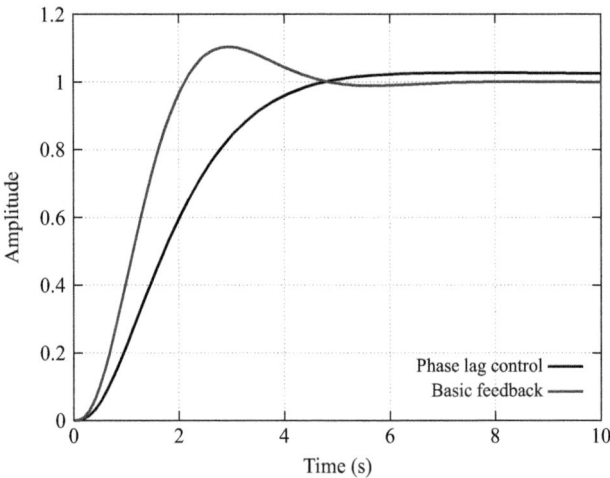

Figure 3.15   Unit step responses of phase lag and basic feedback controlled systems

where k_p is the proportional gain, $k_i \triangleq k_p/T_i$ is the integral gain and T_i is the integral time constant. PI controller is a special case of phase lag compensator with its pole placed at the origin as illustrated in Figure 3.17. PI controller can, thus, be viewed as phase lag compensator with $\beta \to \infty$.

GM = 22.9 dB (at 3.32 rad/s), PM = 70° (at 0.5 rad/s)

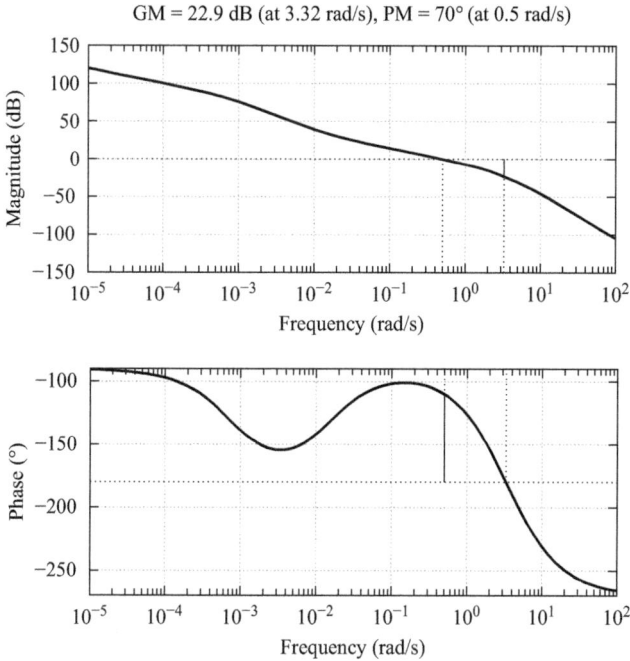

Figure 3.16 Bode plots of phase lag controlled system

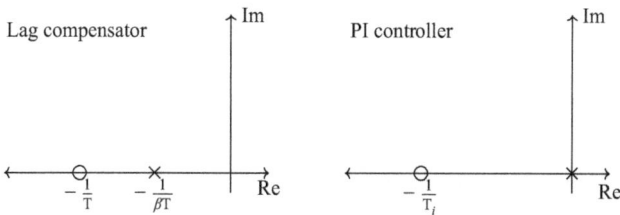

Figure 3.17 The pole-zero locations of phase lag compensator and PI controller

The Bode magnitude and phase angle plots of phase lag and PI controllers are drawn side-by-side in Figure 3.18. The Bode plots are identical at high frequency with differences at lower frequencies. The analytic design procedure for PI control is very similar to that of the phase lag compensator and is presented next.

The plant $G(s)$, provided in (3.3), which consists of a dead-time τ_d and a real rational transfer function part $G_p(s)$. The equation, which should be satisfied for obtaining the desired phase margin ϕ_m at the desired gain crossover frequency ω_{gc} of the open-loop control system $C_{PI}(s)G(s)H(s)$, is

$$\left(\frac{k_i + jk_p\omega_{gc}}{j\omega_{gc}}\right) G_p(j\omega_{gc})H(j\omega_{gc})e^{-j\tau_d\omega_{gc}} = e^{j(-\pi+\phi_m)}. \tag{3.50}$$

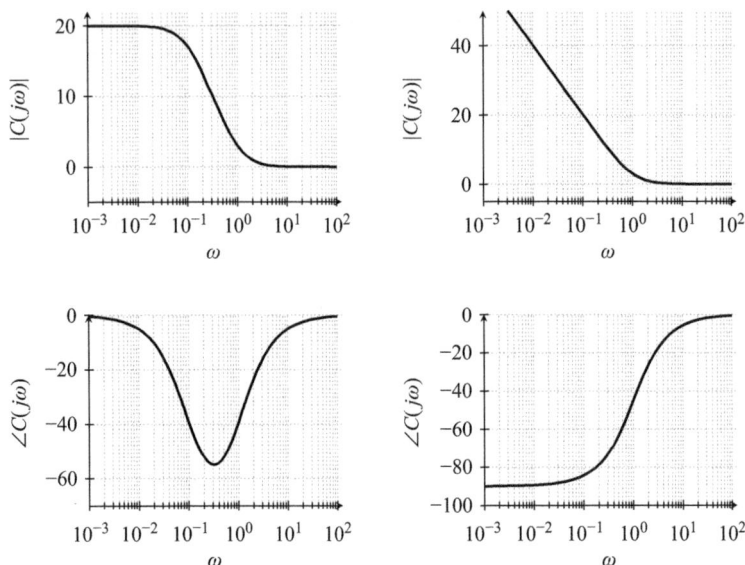

Figure 3.18   Bode plots of phase lag compensator (left) and PI controller (right)

The complex equation (3.50) results in the magnitude and angle equations:

$$\left| \left(\frac{k_i + jk_p\omega_{gc}}{j\omega_{gc}} \right) G_p(j\omega_{gc})H(j\omega_{gc})e^{-j\tau_d\omega_{gc}} \right| = 1, \tag{3.51}$$

$$\theta \triangleq \angle \left(\frac{k_i + jk_p\omega_{gc}}{j\omega_{gc}} \right) = -\pi + \phi_m + \tau\omega_{gc} - \angle G_p(j\omega_{gc})H(j\omega_{gc}). \tag{3.52}$$

Equation (3.51) can be re-written as

$$\left| \frac{k_i + jk_p\omega_{gc}}{j\omega_{gc}} \right| = \frac{1}{|G_p(j\omega_{gc})H(j\omega_{gc})|}, \tag{3.53}$$

which is the gain condition that should be satisfied. Once the desired gain crossover frequency ω_{gc} is chosen, the right-hand side of (3.53) is fixed and can be used to determine its left-hand side, which is the magnitude of the PI controller at the gain crossover frequency. Once this magnitude is obtained, by using (3.53), the PI controller becomes

$$C_{PI}(j\omega_{gc}) = \frac{k_i + jk_p\omega_{gc}}{j\omega_{gc}} = \left| \frac{k_i + jk_p\omega_{gc}}{j\omega_{gc}} \right| e^{j\theta} = \frac{\cos\theta + j\sin\theta}{|G_p(j\omega_{gc})H(j\omega_{gc})|} \tag{3.54}$$

at the gain crossover frequency. Multiplying both sides of (3.54) by $j\omega_{gc}$ and simplifying results in

$$k_i + jk_p\omega_{gc} = -\frac{\omega_{gc}\sin\theta}{|G_p(j\omega_{gc})H(j\omega_{gc})|} + j\frac{\omega_{gc}\cos\theta}{|G_p(j\omega_{gc})H(j\omega_{gc})|}. \tag{3.55}$$

The real and imaginary parts of the complex valued equation, shown in (3.55), are

$$\text{Real:} \quad k_i = -\frac{\omega_{gc} \sin \theta}{|G_p(j\omega_{gc})H(j\omega_{gc})|}, \tag{3.56}$$

$$\text{Imaginary:} \quad k_p\omega_{gc} = \frac{\omega_{gc} \cos \theta}{|G_p(j\omega_{gc})H(j\omega_{gc})|} \tag{3.57}$$

and provide the solution for the integral and proportional gains. The integral time constant can be obtained by using (3.56) and (3.57), and the definition of the PI controller in (3.49) as

$$\frac{1}{T_i} = \frac{k_i}{k_p} = -\omega_{gc} \tan \theta. \tag{3.58}$$

There are two parameters which are the gains k_p and k_i for a PI controller that can be tuned as compared to the three tuneable parameters in a phase lag compensator. Consequently, the analytic design of a PI controller will aim for the achievement of the desired phase margin at the chosen gain crossover frequency and will not involve a static error constant. The steps involved in this procedure is outlined next.

Procedure:

Step 1   Generate the Bode diagram of the open-loop system $G(s)H(s)$, and locate its gain crossover frequency $\tilde{\omega}_{gc}$. Choose the new gain crossover frequency ω_{gc} to be close to the previous one and possibly with $|G(j\omega_{gc})H(j\omega_{gc})|$ having magnitude lower than 0 dB.

Step 2   Evaluate

$$\theta = -\pi + \phi_m + \tau_d\omega_{gc} - \angle G_p(j\omega_{gc})H(j\omega_{gc}),$$

and solve for the PI controller gains using

$$k_i = -\frac{\sin \theta}{|G_p(j\omega_{gc})H(j\omega_{gc})|}\omega_{gc},$$

$$k_p = \frac{\cos \theta}{|G_p(j\omega_{gc})H(j\omega_{gc})|},$$

and

$$\frac{1}{T_i} = \frac{k_i}{k_p} = -\omega_{gc} \tan \theta.$$

Step 3   Draw the Bode plots of the compensated system

$$\left(k_p + \frac{k_i}{s}\right) G(s)H(s).$$

Evaluate the phase margin and gain margin to make sure that the desired phase margin is obtained and that the resulting gain margin is satisfactory. Here, $-90° < \theta < 0°$ should be satisfied for the analytical solution procedure to make sense as that is the phase angle range that can be provided by a PI controller. If this is not satisfied, a different desired gain crossover frequency should be used. Otherwise, a negative value of k_i will result.

Example 3.4 (PI controller design)

Consider the same plant, as given in Example 3.1, which is

$$G(s) = \frac{1}{s(0.2s + 1)(0.45s + 1)}.$$

Let us use the specification of a phase margin that is at least 70° at the desired gain crossover frequency of 0.5 rad/s. A static velocity error constant will not be specified as there are not enough control parameters that can be tuned to achieve that extra specification. By utilising the Conventional Controller Design part of the COMES toolbox, the PI controller that satisfies the desired design specifications is designed as illustrated in Figure 3.19.

The resulting PI controller is obtained as

$$C_{PI}(s) = \frac{k_p s + k_i}{s} = \frac{0.515s + 0.00723}{s}. \tag{3.59}$$

The unit step responses of the uncompensated and PI-controlled system are displayed in Figure 3.20. The PI-controlled system is slower but has very small overshoot as it has a phase margin of 70°. The Bode plots with and without PI control are shown in Figure 3.21. It is verified that the desired phase margin and gain crossover frequency have been obtained. The resulting 22.9 dB gain margin at the phase crossover frequency of 3.32 rad/s that is achieved is satisfactory.

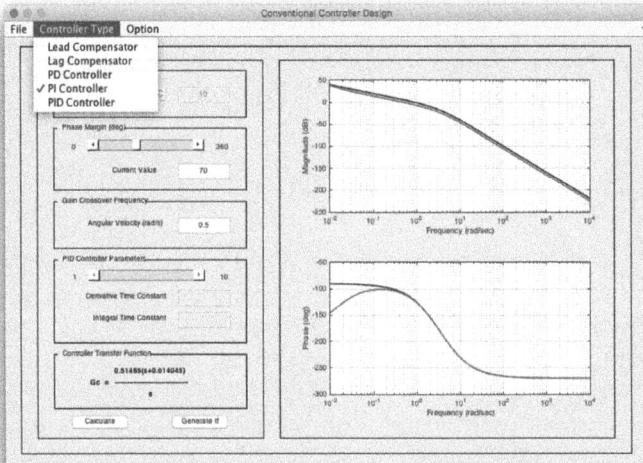

Figure 3.19   PI controller design in COMES toolbox

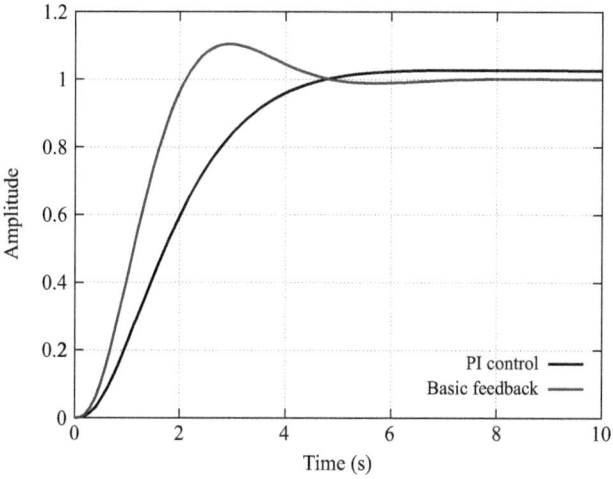

Figure 3.20 Unit step responses of PI and basic feedback controlled systems

GM = 22.9 dB (at 3.32 rad/s), PM = 70° (at 0.5 rad/s)

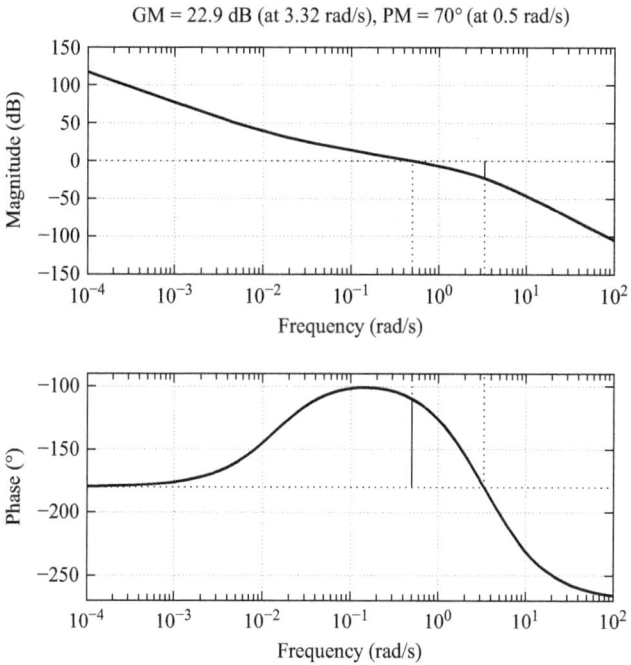

Figure 3.21 Bode plots of PI-controlled system

3.4   Phase lag-lead compensation

The phase lag-lead compensator consists of one gain, two zeros and two poles. Its transfer function is the cascade combination of a phase lag and a phase lead compensator and is given by

$$C_{\text{lag-lead}}(s) = K_c \alpha \beta \underbrace{\left(\frac{T_1 s + 1}{\alpha T_1 s + 1}\right) \left(\frac{T_2 s + 1}{\beta T_2 s + 1}\right)}_{C_{pn}(s)}, \tag{3.60}$$

where $\alpha < 1$ and $\beta > 1$, and

$$C_{pn}(s) = \left(\frac{T_1 s + 1}{\alpha T_1 s + 1}\right)\left(\frac{T_2 s + 1}{\beta T_2 s + 1}\right) \tag{3.61}$$

is the unity d.c. gain part of the phase lag-lead controller. The expression

$$\frac{s + \frac{1}{T_1}}{s + \frac{1}{\alpha T_1}} \tag{3.62}$$

is the phase lead compensation part while the expression

$$\frac{s + \frac{1}{T_2}}{s + \frac{1}{\beta T_2}} \tag{3.63}$$

is the phase lag compensation part. The polar plot of the phase lag-lead compensator in Figure 3.22 shows that it introduces both negative (undesired) and positive (desired) phase angle to the system. This can be seen more clearly in the Bode plots of Figure 3.23. The phase lead portion of the phase lag-lead compensator (the portion involving T_1) alters the frequency response curve by adding phase lead angle and increasing the phase margin at the gain crossover frequency. The phase lag portion

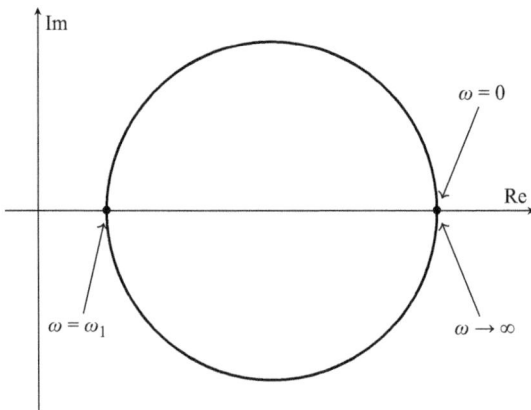

Figure 3.22   Polar plot of a phase lead-lag compensator

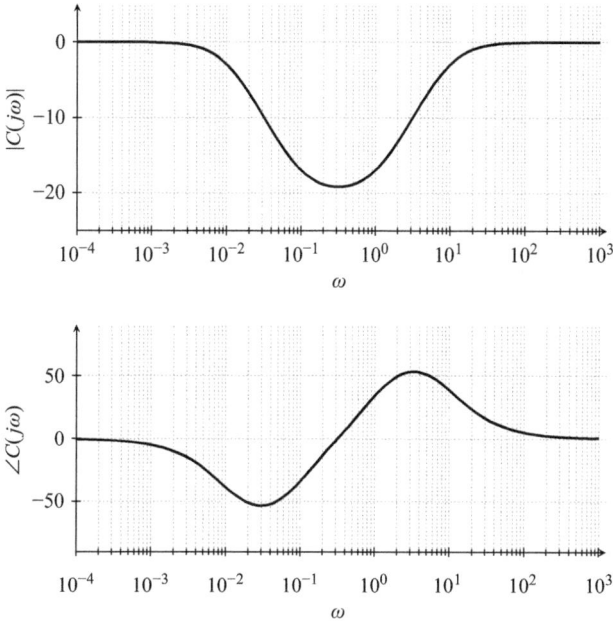

Figure 3.23 Bode plots of phase lead-lag compensator with $K_c = 1$, $\alpha = 0.1$, $\beta = 10$, $T_1 = 1$ and $T_2 = 10$

(the portion involving T_2) enables the increase of gain at the low-frequency range to improve the steady-state error performance.

The approach recommended in this book for phase lag-lead control is to first design a phase lead compensator and then design a phase lead compensator for the phase lead compensated plant. The formulas and COMES toolbox tools presented in previous sections can be used to achieve that purpose easily. For the sake of completeness, however, the analytical design equations for the phase lag-lead controller are presented below.

The equation that should be satisfied for obtaining the desired phase margin ϕ_m at the chosen gain crossover frequency ω_{gc} is

$$K_c \alpha \beta C_{pn}(j\omega_{gc})G_p(j\omega_{gc})H(j\omega_{gc})e^{-j\tau_d\omega_{gc}} = e^{j(-\pi+\phi_m)}. \qquad (3.64)$$

The complex equation (3.64) results in the following magnitude and angle equations:

$$|K_c \alpha \beta C_{pn}(j\omega_{gc})G_p(j\omega_{gc})H(j\omega_{gc})| = 1, \qquad (3.65)$$

$$\theta \triangleq \angle C_{pn}(j\omega_{gc}) = -\pi + \phi_m + \tau_d\omega_{gc} - \angle G_p(j\omega_{gc})H(j\omega_{gc}). \qquad (3.66)$$

Equation (3.65) can be re-written as

$$|C_{pn}(j\omega_{gc})| = \frac{1}{K_c \alpha \beta |G_p(j\omega_{gc})H(j\omega_{gc})|}, \qquad (3.67)$$

which is the gain condition that should be satisfied. Once the desired gain crossover frequency ω_{gc} is chosen and once $K_c\alpha\beta$ is selected to satisfy a static error constant, the right-hand side of (3.67) is fixed and can be used to determine its left-hand side which is the magnitude of the unity d.c. gain normalised phase lag-lead compensator at the gain crossover frequency. Once this magnitude is obtained, the normalised phase lag-lead compensator becomes

$$C_{pn}(j\omega_{gc}) = |C_{pn}(j\omega_{gc})|e^{j\theta} \tag{3.68}$$

at the gain crossover frequency. By using (3.61), (3.67) and (3.68),

$$\left(\frac{1+jT_1\omega_{gc}}{1+j\alpha T_1\omega_{gc}}\right)\left(\frac{1+jT_2\omega_{gc}}{1+j\beta T_2\omega_{gc}}\right) = \frac{\cos\theta + j\sin\theta}{K_c\alpha\beta|G_p(j\omega_{gc})H(j\omega_{gc})|} \tag{3.69}$$

is obtained. Using a common denominator on both sides of (3.69) and equating the resulting numerators results in

$$K_c\alpha\beta|G_p(j\omega_{gc})H(j\omega_{gc})|(1+jT_1\omega_{gc})(1+jT_2\omega_{gc})$$
$$= (1+j\alpha T_1\omega_{gc})(1+j\beta T_2\omega_{gc})(\cos\theta + j\sin\theta), \tag{3.70}$$

which can be re-expressed as

$$K_c\alpha\beta|G_p(j\omega_{gc})H(j\omega_{gc})|\left(-T_1T_2\omega_{gc}^2 + 1 + j\omega_{gc}(T_1+T_2)\right)$$
$$= \left((1-\alpha\beta T_1T_2\omega_{gc}^2)\cos\theta - (\alpha T_1+\beta T_2)\omega_{gc}\sin\theta\right)$$
$$+ j\left((1-\alpha\beta T_1T_2\omega_{gc}^2)\sin\theta + (\alpha T_1+\beta T_2)\omega_{gc}\cos\theta\right). \tag{3.71}$$

The real and imaginary parts of the complex valued equation, seen in (3.71), are

$$\text{Real}: \quad \cos\theta - (\alpha T_1+\beta T_2)\omega_{gc}\sin\theta - \alpha\beta T_1T_2\omega_{gc}^2\cos\theta$$
$$= K_c\alpha\beta|G_p(j\omega_{gc})H(j\omega_{gc})|(1-T_1T_2\omega_{gc}^2), \tag{3.72}$$

$$\text{Imaginary}: \quad \sin\theta + (\alpha T_1+\beta T_2)\omega_{gc}\cos\theta - \alpha\beta T_1T_2\omega_{gc}^2\sin\theta$$
$$= K_c\alpha\beta|G_p(j\omega_{gc})H(j\omega_{gc})|(T_1+T_2)\omega_{gc}. \tag{3.73}$$

Equations (3.72) and (3.73) have to be solved to obtain phase lead compensator's pole and zero locations as chosen two parameters from the four possibilities of T_1, αT_1, T_2 and βT_2. Despite doing this analytically appears to be a difficult task, it is easy to solve (3.72) and (3.73) numerically in MATLAB.

3.4.1   PID control as a special case of phase lag-lead compensation

The proportional-integral-derivative (PID) control is given by

$$C_{PID}(s) = \frac{k_p}{s}\left(T_d s^2 + s + \frac{1}{T_i}\right) = \frac{k_d s^2 + k_p s + k_i}{s}, \tag{3.74}$$

where the gains and time constants have the same definitions as in the PD and PI control cases treated in previous sections. Its frequency response is given by

$$C_{\text{PID}}(j\omega) = \frac{k_i - k_d\omega^2 + jk_p\omega}{j\omega} = k_p + j\left(k_d\omega - \frac{k_i}{\omega}\right). \tag{3.75}$$

The PID controller is a special case of phase lag-lead control with the two compensator poles placed at the origin and infinity. The analytical solution procedure in the frequency-domain is applied to PID control in this section.

The plant $G(s)$ is given by (3.3) which consists of a dead-time τ_d and a real rational transfer function part $G_p(s)$. The equation that should be satisfied for obtaining the desired phase margin ϕ_m at the selected gain crossover frequency ω_{gc} of the open-loop control system $C_{\text{PID}}(s)G(s)H(s)$ is

$$\left[k + j\left(k_d\omega_{gc} - \frac{k_i}{\omega_{gc}}\right)\right] G_p(j\omega_{gc})H(j\omega_{gc})e^{-j\tau_d\omega_{gc}} = e^{j(-\pi+\phi_m)}. \tag{3.76}$$

The complex equation (3.76) results in the magnitude and angle equations:

$$\left|\left[k_p + j\left(k_d\omega_{gc} - \frac{k_i}{\omega_{gc}}\right)\right] G_p(j\omega_{gc})H(j\omega_{gc})\right| = 1, \tag{3.77}$$

$$\theta \triangleq \angle\left[k_p + j\left(k_d\omega_{gc} - \frac{k_i}{\omega_{gc}}\right)\right]$$
$$= -\pi + \phi_m + \tau_d\omega_{gc} - \angle G_p(j\omega_{gc})H(j\omega_{gc}). \tag{3.78}$$

Equation (3.77) can be re-written as

$$\left|k_p + j\left(k_d\omega_{gc} - \frac{k_i}{\omega_{gc}}\right)\right| = \frac{1}{|G_p(j\omega_{gc})H(j\omega_{gc})|}, \tag{3.79}$$

which is the gain condition that should be satisfied. Once the desired gain crossover frequency ω_{gc} is chosen, the right-hand side of (3.79) is fixed and can be used to determine its left-hand side, which is the magnitude of the PID controller at the gain crossover frequency. Once this magnitude is obtained, using (3.75) the PID controller becomes

$$C_{\text{PID}}(j\omega_{gc}) = \left|k_p + j\left(k_d\omega_{gc} - \frac{k_i}{\omega_{gc}}\right)\right| e^{j\theta} = \frac{\cos\theta + j\sin\theta}{|G_p(j\omega_{gc})H(j\omega_{gc})|} \tag{3.80}$$

at the gain crossover frequency. Using (3.75) and (3.79),

$$k_p + j\left(k_d\omega_{gc} - \frac{k_i}{\omega_{gc}}\right) = \frac{\cos\theta + j\sin\theta}{|G_p(j\omega_{gc})H(j\omega_{gc})|} \tag{3.81}$$

is obtained. The real and imaginary parts of the complex valued equation, seen in (3.81), are

$$\text{Real:} \quad k_p = \frac{\cos\theta}{|G_p(j\omega_{gc})H(j\omega_{gc})|}, \tag{3.82}$$

$$\text{Imaginary:} \quad k_d\omega_{gc} - \frac{k_i}{\omega_{gc}} = \frac{\sin\theta}{|G_p(j\omega_{gc})H(j\omega_{gc})|}, \tag{3.83}$$

and provide the solution for two out of the proportional, derivative and integral gains. The proportional gain k_p and one of the gains (k_d and k_i) can be tuned using (3.82) and (3.83). The user should fix either the derivative or the integral gain and use the formulas (3.82) and (3.83) to achieve a desired phase margin at the selected gain crossover frequency. It may be possible to choose the controller gain not selected for tuning to satisfy a desired static error constant requirement. The steps involved in the frequency domain analytical solution procedure are outlined next.

Procedure:

Step 1 Generate the Bode diagram of the open-loop system $G(s)H(s)$ and locate its gain crossover frequency $\tilde{\omega}_{gc}$. Choose the new gain crossover frequency ω_{gc}.

Step 2 Evaluate

$$\theta = -\pi + \phi_m + \tau_d \omega_{gc} - \angle G_p(j\omega_{gc})H(j\omega_{gc})$$

and solve for the PID controller gains k_p and one of k_d and k_i using

$$k_p = \frac{\cos\theta}{|G_p(j\omega_{gc})H(j\omega_{gc})|},$$

$$k_d \omega_{gc} - \frac{k_i}{\omega_{gc}} = \frac{\sin\theta}{|G_p(j\omega_{gc})H(j\omega_{gc})|}.$$

Step 3 Draw the Bode plots of the compensated system

$$\left(k_p + k_d s + \frac{k_i}{s}\right) G(s)H(s).$$

Evaluate the phase margin and gain margin to make sure that the desired phase margin is obtained and that the resulting gain margin is satisfactory. Here, $-90° < \theta < 90°$ should be satisfied for the analytical solution procedure to make sense as that is the phase angle range that can be provided by a PID controller. If this is not satisfied, a different desired gain crossover frequency should be used.

Example 3.5 (PID controller design)

Consider the same plant, as given in Example 3.1, which is

$$G(s) = \frac{1}{s(0.2s + 1)(0.45s + 1)}.$$

Let us use the specification of a phase margin that is at least 70° at the desired gain crossover frequency of 4 rad/s. Since we can only choose two of the three PID controller parameters, the derivative time constant will be chosen as 0.5 s. The procedure developed in this subsection can then be used to determine the proportional and integral PID gains. By utilising the Conventional Controller Design part of the COMES toolbox, the PID controller that satisfies the desired design specifications is designed as illustrated in Figure 3.24.

The resulting PID controller is obtained as

$$C_{\text{PID}}(s) = \frac{k_d s^2 + k_p s + k_i}{s} = \frac{2.964 s^2 + 1.903 s + 5.928}{s}. \tag{3.84}$$

The unit step responses of the uncompensated and PID-controlled systems are displayed in Figure 3.25. The PID-controlled system is faster, but it has considerable overshoot and exhibits an oscillatory behaviour, which is not typical for a phase margin of 70°. The Bode plots with and without PID control are shown in Figure 3.26. It is verified that the desired phase margin and gain crossover frequency have been obtained. The resulting 22.9 dB gain margin at the phase crossover frequency of 3.32 rad/s that is achieved is satisfactory.

This PID controller is robust as it has high phase and gain margins. This is one problem with robust controllers as they may not always have the best nominal performance. Let us repeat the design with different specifications to improve the transient response over what is displayed in Figure 3.25. Let us keep the desired phase margin at 70°, i.e., still a robust controller, and increase the derivative time constant to 1 s while reducing the gain crossover frequency to 1.5 rad/s. Using the COMES toolbox, this PID controller becomes

$$C_{\text{PID}}(s) = \frac{k_d s^2 + k_p s + k_i}{s} = \frac{1.158 s^2 + 1.624 s + 1.158}{s}, \tag{3.85}$$

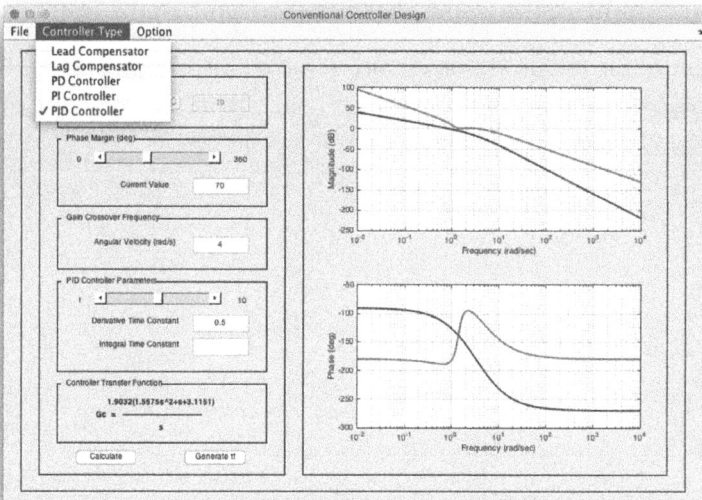

Figure 3.24   PID controller design in COMES toolbox

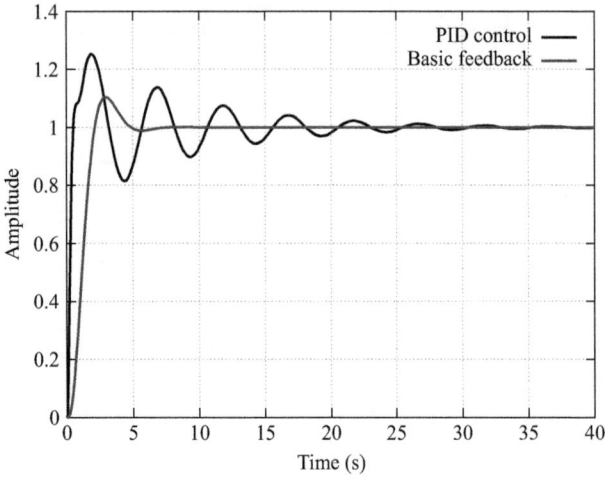

Figure 3.25   Unit step responses of PID and basic feedback controlled systems

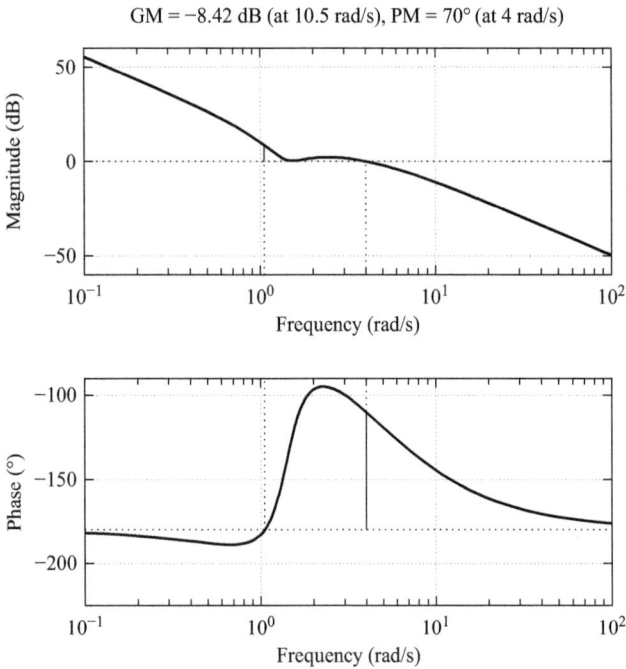

Figure 3.26   Bode plots of PID controlled system

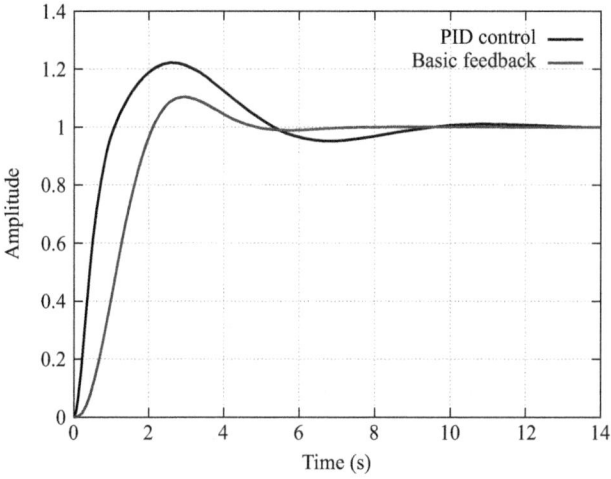

Figure 3.27   Unit step responses of PID and basic feedback controlled systems

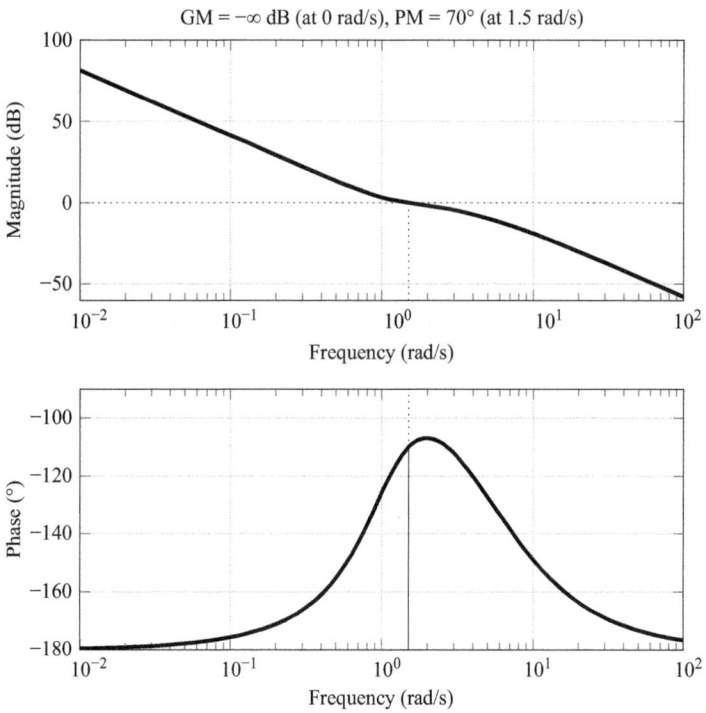

Figure 3.28   Bode plots of PID-controlled system

and its step response and Bode plots are displayed in Figures 3.27 and 3.28, respectively. The transient response is better than before, the desired phase margin and gain crossover frequency are obtained and the gain margin is infinite. Nominal performance is still not so good as there is larger overshoot even as compared to basic feedback due to the larger overshoot and slightly more oscillatory response even though the rise time is smaller.

3.5 Optimisation-based conventional controller design in MATLAB and Simulink

If one is only interested in nominal performance and a robust controller design in the form of the desired phase margin or a parameter space based approach is not vital, numerical optimisation based tools available in computer aided control system analysis and design software like MATLAB and Simulink can also be used. These numerical optimisation based PID or controller parameter tuners use an iterative approach. It is usually a difficult task to determine the algorithm used inside the tuner and how changes may be made. The use of the PID tuner, which is a part of the PID block in Simulink, is illustrated with an example next.

Example 3.6

Consider the same plant as provided in Examples 3.1–3.5, which is

$$G(s) = \frac{1}{s(0.2s+1)(0.45s+1)}.$$

The standard Simulink block diagram shown in Figure 3.29 is constructed first. The PID block in Figure 3.29 is double clicked to enter parameters of the last PID design in Example 3.5 as illustrated in Figure 3.30. N in Figure 3.30 is the cut-off frequency of the limited differentiator that needs to be used to make the PID controller causal and hence realisable. It has been set at a large value deliberately to reduce its effect on the standard PID. Once the Tune button is pressed the Simulink Control Design option works by first linearising the system and choosing optimal PID parameters.

The numerical optimisation tuning results in the tuned PID given by

$$C_{\text{PID}}(s) = k_p + \frac{k_i}{s} + k_d s \frac{N}{s+N} = 3.554 + \frac{0.9312}{s} + 1.960s\frac{381}{s+381}.$$

The tuned response and the initial (block) response are displayed in Figure 3.31 to demonstrate the improvement in performance. The controller parameters and performance and robustness related parameters before (block) and after tuning (tuned) are also displayed as a table. It is possible to experiment with the trade-off between response time and robustness to a limited extent by using sliders in the design screen in Figure 3.31. An analysis of the results in Figure 3.31 shows that while the tuned response has better transient properties, its robustness as measured by the phase margin has decreased. It is not possible to directly design for and obtain specified robustness measures like the desired phase margin and a desired gain crossover frequency with this method. As numerical optimisation is used for tuning,

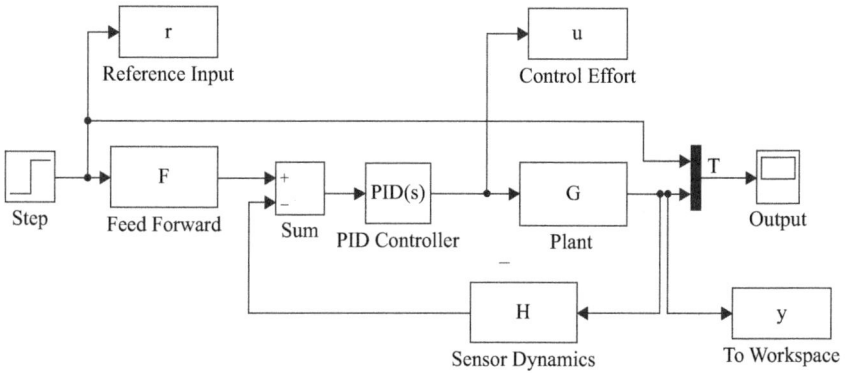

Figure 3.29   Simulink block diagram with PID block

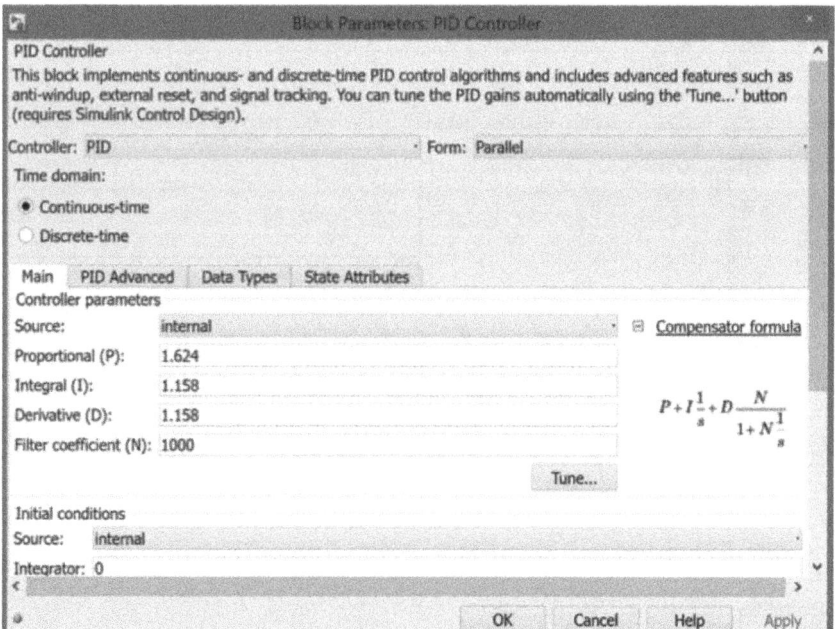

Figure 3.30   PID block parameters in Simulink

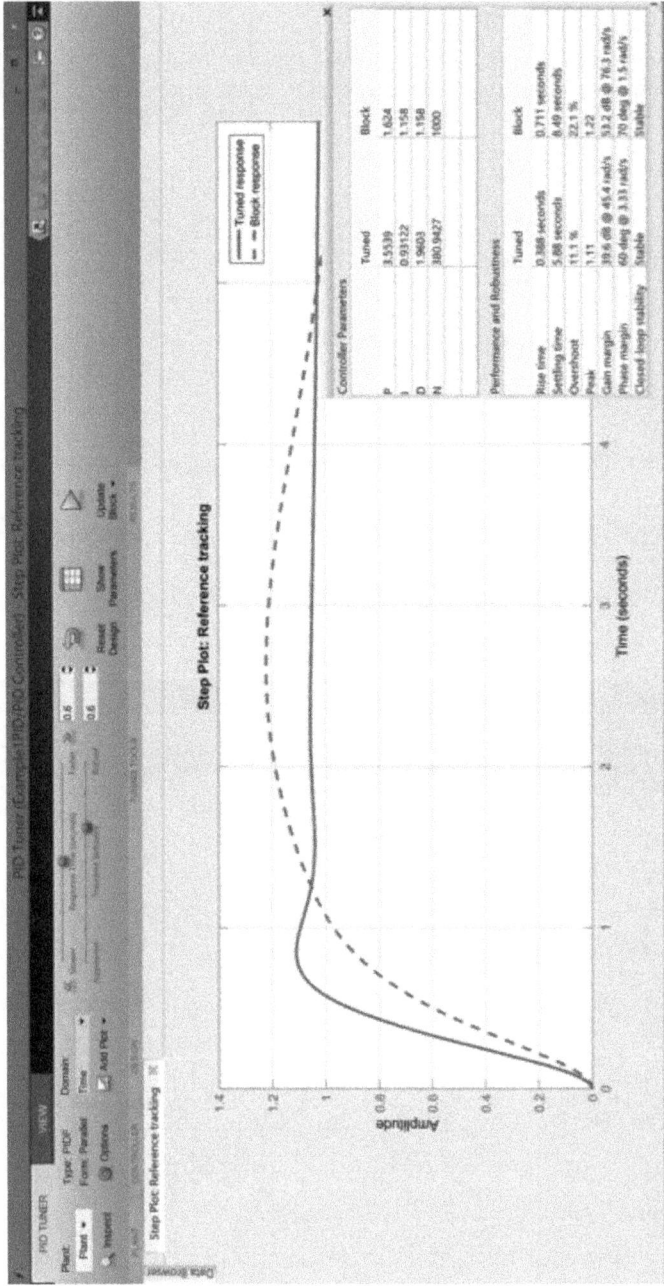

Figure 3.31   Comparison of tuned PID and initial PID step responses

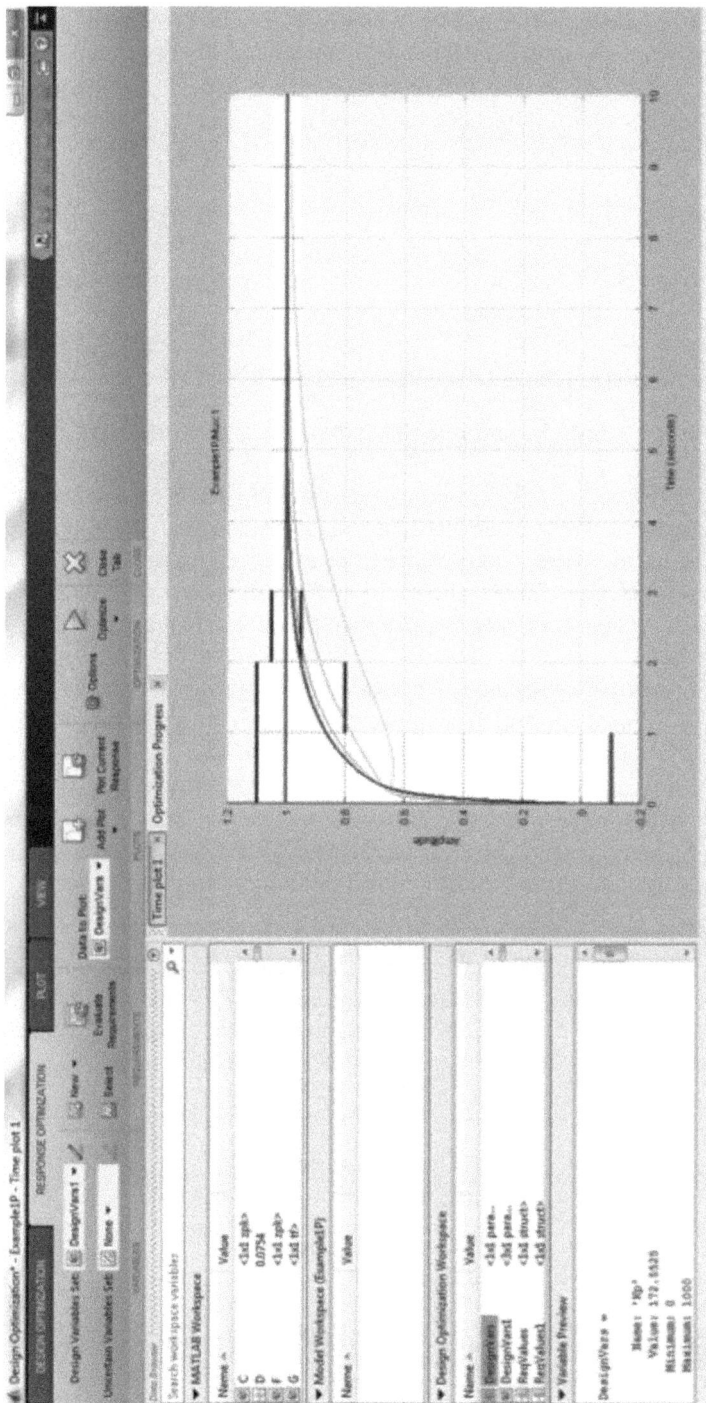

Figure 3.32 Numerical optimisation based tuning for desired step response shape

a large number of simulations have to be run at each iteration of the optimisation, resulting in a large time for the overall tuning when the plant model is complicated. This is not an issue with the analytical solution approach, presented earlier in this chapter, while obtaining the frequency response of the plant at a single design frequency is computationally very simple and fast. Numerical optimisation based tuning is an effective approach to quickly obtain a PID controller with a satisfactory transient response for relatively simple plants like the example plant considered until this point. These controllers can be used for benchmarking purposes. The time response can also be optimised directly in MATLAB/Simulink as illustrated in Figure 3.32 for a different plant.

3.6 Parameter space based robust conventional controller design

Parameter space methods for both controller design and uncertainty analysis were presented in Chapter 2. These parameter space based methods are very easily applicable to solving for two chosen controller or plant uncertainty parameters of a plant under conventional control. These two chosen parameters may be two of the gains in phase lag-lead or PID controllers or two uncertain parameters in the plant model for known controllers.

This section has two subsections. Subsection 3.6.1 compares the analytical solution approach developed and presented in this chapter with the phase margin bound determination based parameter space computations of Chapter 2. Subsection 3.6.2 is on $\mathscr{D}$-stability based parameter space design of a PID for steering control of a road vehicle in automated path following.

3.6.1 *Comparison of analytical approach and parameter space phase margin bound computations*

As a specific example, consider the analytical equations (3.66), (3.72), and (3.73) obtained in the previous section for analytical design of PID controller gains for satisfying a desired phase margin ϕ_m at a chosen gain crossover frequency ω_{gc}. Let us use the frequency ω as a free parameter in these design equations instead of a fixed frequency ω_{gc} as

$$k_p = \frac{\cos\left(-\pi + \phi_m + \tau_d\omega - \angle G_p(j\omega)H(j\omega)\right)}{|G_p(j\omega)H(j\omega)|}, \tag{3.86}$$

$$k_p\omega - \frac{k_i}{\omega} = \frac{\sin\left(-\pi + \phi_m + \tau_d\omega - \angle G_p(j\omega)H(j\omega)\right)}{|G_p(j\omega)H(j\omega)|}. \tag{3.87}$$

Equations (3.86) and (3.87) can be evaluated for a grid of frequency points to obtain the parameter space plot of PID gains that satisfy a specified phase margin ϕ_m on the boundary. The gain crossover frequency cannot be specified in this approach

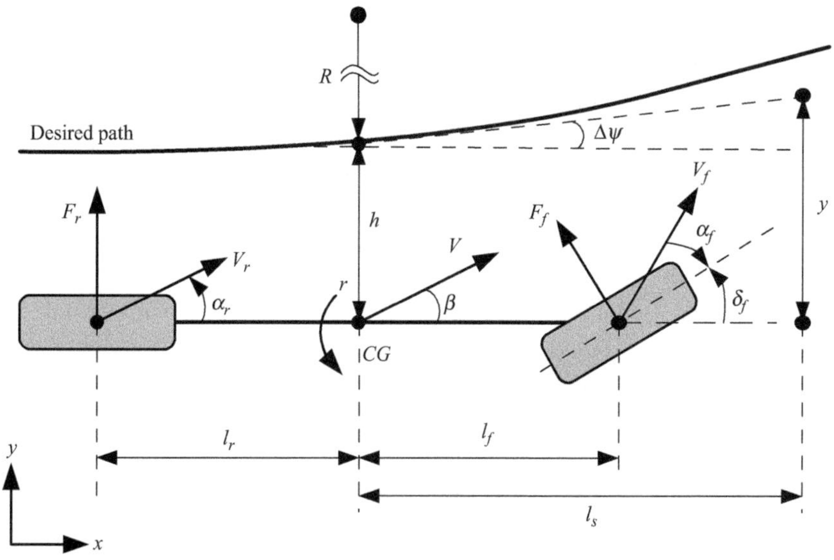

Figure 3.33 Vehicle and path geometry

as frequency ω is a free parameter and changes over the boundary obtained. As in Chapter 2, regions where the phase margin bound of ϕ_m being less than a certain value are satisfied will be inside or outside or to one side or the other side of the boundary. A similar approach can be used for all of the phase lead, phase lag, phase lag-lead, PD and PI controllers treated in this chapter.

3.6.2 Case study of parameter space based conventional controller design

This case study on automated path following is taken from [9]. The diagram in Figure 3.33 shows the vehicle and path geometry along with some relevant parameters. A single track vehicle model is used as the basic underlying steering dynamic model. The linearised vehicle model used for robust path following controller design is

$$\begin{bmatrix} \dot{\beta} \\ \dot{r} \\ \Delta\dot{\psi} \\ \dot{y} \end{bmatrix} = \begin{bmatrix} a_{11} & a_{12} & 0 & 0 \\ a_{21} & a_{22} & 0 & 0 \\ 0 & 1 & 0 & 0 \\ V & \ell_s & V & 0 \end{bmatrix} \begin{bmatrix} \beta \\ r \\ \Delta\psi \\ y \end{bmatrix} + \begin{bmatrix} b_{11} & 0 \\ b_{21} & 0 \\ 0 & -V \\ 0 & 0 \end{bmatrix} \begin{bmatrix} \delta_f \\ \rho_{ref} \end{bmatrix}, \quad (3.88)$$

where β, r, V, $\Delta\psi$, ℓ_s and y are vehicle side slip angle, vehicle yaw rate, vehicle velocity, yaw angle relative to the desired path's tangent, the preview distance and lateral deviation from the desired path at the preview distance, respectively. The

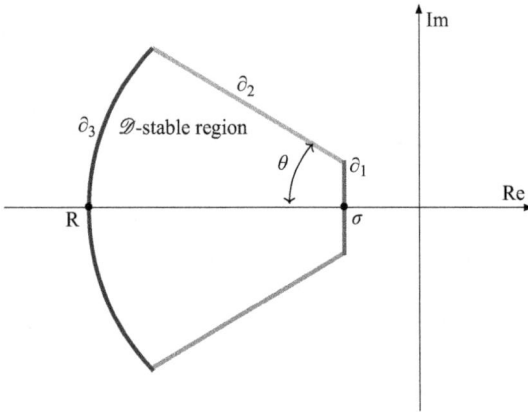

Figure 3.34 $\mathscr{D}$-stable region in the complex plane

control input is the steering angle δ_f; $\rho_{\text{ref}} = 1/R$ is the road curvature where R is the road radius of curvature. The remaining terms are

$$a_{11} = -\frac{c_r + c_f}{\tilde{m}V},$$

$$a_{12} = -1 + \frac{c_r l_r - c_f l_f}{\tilde{m}V^2},$$

$$a_{21} = \frac{c_r l_r - c_f l_f}{\tilde{J}},$$

$$a_{22} = -\frac{c_r l_r^2 + c_f l_f^2}{\tilde{J}V^2},$$

$$b_{11} = \frac{c_f}{\tilde{m}V},$$

$$b_{21} = \frac{c_f l_f}{\tilde{J}},$$

where $\tilde{m} \triangleq m/\mu$ is the virtual mass, $\tilde{J} \triangleq J/\mu$ is the virtual moment of inertia, μ is the road friction coefficient, m is the vehicle mass, J is the moment of inertia, c_f and c_r are the cornering stiffnesses, l_f is the distance from the centre of gravity of the vehicle (CG) to the front axle and l_r is the distance from the CG to the rear axle. The values of the parameters used are taken from [9]. The vehicle mass, the vehicle velocity and the road friction coefficient are taken as uncertain parameters.

A robust PID path following steering controller is designed based on the parameter space approach here. Two parameters of the PID controller are selected as free design parameters. The solution regions, which satisfy the $\mathscr{D}$-stability requirements shown graphically in Figure 3.34 are calculated and plotted, based on the parameter space approach in the $k_p - k_d$ parameter space for a chosen k_i value. By repeating this procedure for several k_i values, the three-dimensional controller parameter space

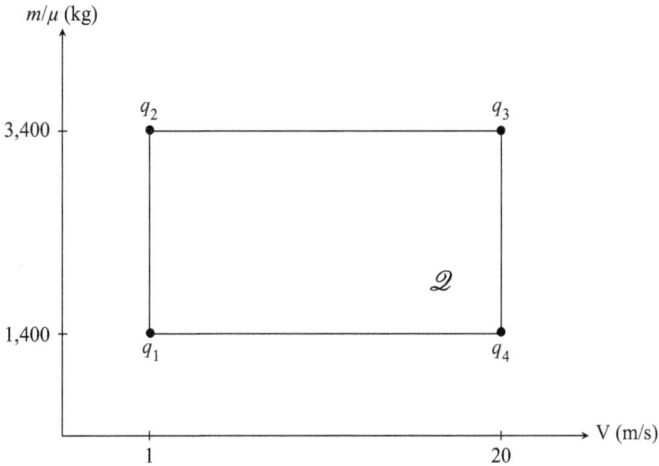

Figure 3.35 Uncertainty box

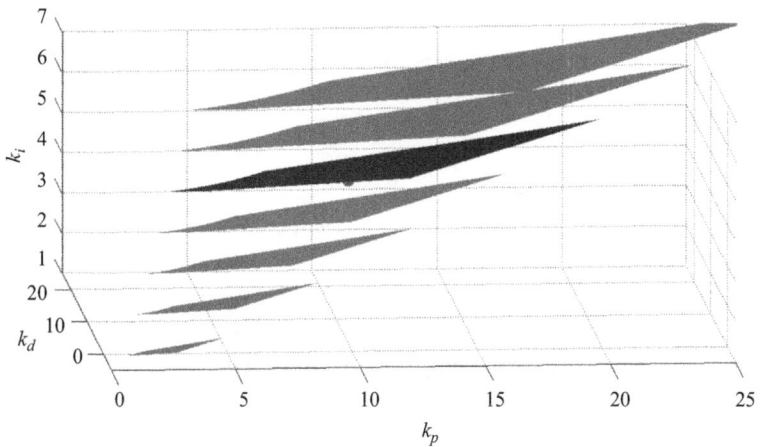

Figure 3.36 Overall $\mathscr{D}$-stability solution regions scheduled by k_i values

of PID gains k_p, k_i and k_d is obtained. The $\mathscr{D}$-stability boundaries, illustrated in Figure 3.34, are formed by assuming roots no closer than 0.5 to the imaginary axis and no further in magnitude than 2.7 from the imaginary axis ($\sigma = 0.5$ and $R = 2.7$). A minimum damping ratio corresponding to $\theta = 45°$ is determined as $\sqrt{2}/2$ as a design requirement.

The vehicle mass, the vehicle velocity and the road friction coefficient are taken as uncertain parameters for the calculations. The uncertain parameter ranges are selected as: $m \in [1,400, 1,700]$ (kg) (the nominal value of mass is $1,550$ kg), $\mu \in [0.5, 1]$ and $V \in [1, 20]$ (m/s), respectively. The virtual mass, then, is within the range $\tilde{m} = {}^m/_\mu \in$

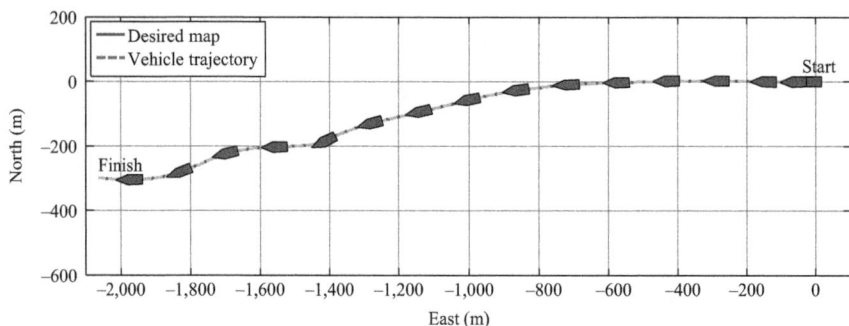

Figure 3.37   Desired path and stroboscopic vehicle trajectory

[1, 400, 3, 400] (kg). The corresponding uncertainty box of virtual mass and vehicle speed is shown in Figure 3.35.

The overall solution region, which combines all solutions for the vertices of the uncertainty box in Figure 3.35, is shown in Figure 3.36. The solution point for the PID controller gains is marked with a dot and its $k_p - k_i$ parameter plane solution is shown with different shading in Figure 3.36. The chosen PID controller is given by

$$C_{\text{PID}}(s) = k_p + \frac{k_i}{s} + k_d s = 10 + \frac{5}{s} + 3s. \tag{3.89}$$

The simulation result in Figure 3.37 shows that this parameter space based PID steering controller does a very good job of following actual road trajectories.

3.7   Case study: Quanser QUBE™ Servo system

We will re-visit the Quanser QUBE™ servo rotational speed control problem that was considered in Section 2.5. The plant model from control input to motor rotational position is given by

$$G(s) = \frac{28}{s(0.1s + 1)}, \tag{3.90}$$

where the numerical values used were rounded to the nearest digit. A phase lead compensator is designed for rotational position control of this rotational servo using the COMES toolbox. The specifications used in the design are a velocity error constant of $K_v = 28$ (no change) a 0 dB bandwidth (gain crossover frequency) of 314 rad/s (100 Hz) and a phase margin of 90°. The resulting phase lead compensator is

$$C_{\text{lead}}(s) = \frac{1.122s + 1}{9.238 \times 10^{-5}s + 1}. \tag{3.91}$$

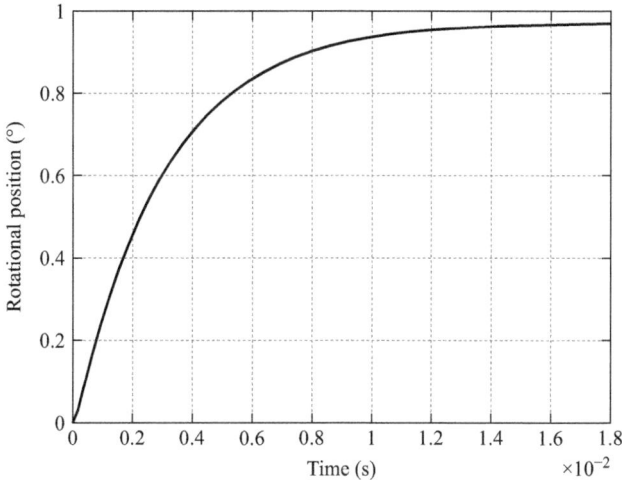

Figure 3.38 Step response of phase lead compensated Quanser QUBE™ servo system's rotational position

Note that this phase lead compensator is almost a PD controller as the coefficient of the leading term in the denominator is almost zero. The step response of the phase lead compensated system in Figure 3.38 shows a very fast response without any overshoot.

3.8 Chapter summary and concluding remarks

We derived an analytical method to easily calculate the parameters of conventional control systems in this chapter. The analytical solution procedure is based on finding formulas for controller gains to satisfy the desired phase margin at the chosen gain crossover frequency. Note that a similar approach can easily be applied to finding formulas for controller gains to achieve a desired gain margin at the selected phase crossover frequency. As designers usually prefer to specify phase margin and bandwidth requirements, the following calculation was not a part of this chapter. Optimisation-based tuners available for tuning controllers in MATLAB and PID tuners in Simulink were also presented as these can be computed very fast if the plant model is known and not very complicated. These optimisation-based controllers can be used for benchmarking purposes. The optimisation may not converge in some cases while in other cases it may not be possible to meet the design specification.

Note that the analytical design method can handle plants with time delay easily as computations are done in the frequency domain. Parameter space methods with frequency domain bounds can also easily deal with a time delay in the plant. Hurwitz stability and $\mathscr{D}$-stability parameter space mapping cannot handle time delays. In those cases, a Pade approximation of appropriate order is used instead of the time delay for these methods to work. The parameter space method for conventional controller design was introduced using a case study. The following chapters will concentrate on other control architectures.

References

[1]   K. Ogata, *Modern Control Engineering*. New York: Prentice Hall, 1990.

[2]   W. Wakeland, "Analytic technique for root locus compensation with quadratic factors," *IEEE Transactions on Automatic Control*, vol. 12, pp. 631–632, 1967.

[3]   ——, "Bode compensator design," *IEEE Transactions on Automatic Control*, vol. 21, pp. 771–773, Oct. 1975.

[4]   G. Thaler and R. Brown, *Servomechanism analysis*. New York: McGraw-Hill, 1953.

[5]   V. D. Toro and S. Parker, *Principles of control systems engineering*. New York, USA: McGraw-Hill, 1960.

[6]   B. Watkins, *Introduction to control systems*. New York: Macmillan, 1969.

[7]   J. R. Mitchell, "Comments on "Bode compensator design"," *IEEE Transactions on Automatic Control*, vol. 22, pp. 869–870, 1977.

[8]   K. S. Yeung, K. W. Wong, and K.-L. Chen, "A non-trial-and-error method for lag-leag compensator design," *IEEE Transactions on Education*, vol. 41, no. 1, pp. 76–80, Feb. 1998.

[9]   M. T. Emirler, H. Wang, B. Aksun-Güvenç, and L. Güvenç, "Automated robust path following control based on calculation of lateral deviation and yaw angle error," in *ASME 2015 Dynamic Systems and Control Conference*, 2015.

Chapter 4

Input shaping control

This chapter is on input shaping, also known as preview control. This is a feedforward control approach that can be applied to reference and disturbance inputs that are known in advance. A discrete-time closed-loop controlled system is considered and its approximate inverse filter that is not causal is designed to achieve desirable input-to-output behaviour. The external input is assumed to be known in advance. Thus, the approximate inverse filter also known as the preview filter as it is not causal. It is applied to the known input offline, resulting in a new input which is applied online to the closed-loop system (hence the name input shaping). This chapter starts with a characterisation of non-minimum phase zeros and approximate inversion methods used to treat them as they cannot be inverted directly. The ZP, ZPG, ZPGE and ZPGO and TSA methods are presented in this chapter as different possible approximate inverse filters for feedforward control.

4.1 Introduction to input shaping control

The aim in reference feedforward tracking control is to obtain a low-pass filter of desired tracking bandwidth between the input and the output to reduce tracking error [1]. A two-degree-of-freedom controller architecture is used to achieve this purpose by cascading the closed-loop system, already under feedback control, with its approximate inverse, as shown in Figure 4.1. The second degree of design freedom afforded by the feedforward control block $\tilde{G}_{fb}^{-1}$ is used to reduce tracking error below what is achievable with only the feedback controller C. While the feedback controller must be limited to causal filters, the approximate inverse filter $\tilde{G}_{fb}^{-1}$ is allowed to be non-causal since off-line computation and hence preview of the input signal R is possible in feedforward tracking control.

Seeking an exact inverse based on pole-zero cancellation to achieve perfect compensation of the plant or closed-loop system is the ideal goal. This is not possible since non-minimum phase (NMP) zeros, if they exist, cannot be inverted directly. NMP zeros are quite common in models of process control systems with transport lag [2] and flexible structures [3]. It is also a well-known fact that even the zero-order-hold equivalent of a minimum phase continuous-time plant with relative degree greater than two will always be NMP for large enough sampling rate [4]. Due to the presence of high-frequency modelling errors, exact inversion is not a good idea even for

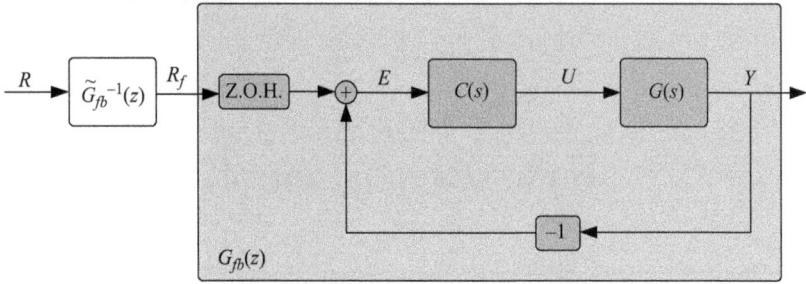

Figure 4.1   Discrete-time reference feedforward tracking control

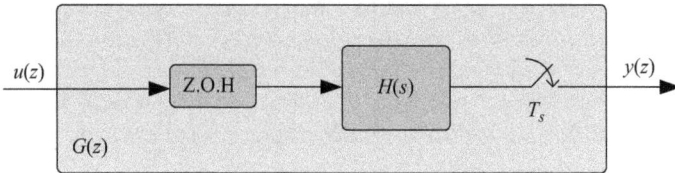

Figure 4.2   ZOH discrete-time equivalent

minimum phase systems and an approximate inverse should be used to achieve low-pass filter characteristics between system input and output for better performance robustness.

4.2   Discrete-time NMP zeros

Consider the zero-order-hold (ZOH) discretised version $G(z)$ of a continuous-time plant $H(s)$ as illustrated in Figure 4.2. For $H(s)$ with n poles and m zeros, $G(z)$ has n poles and $n - 1$ zeros, some of which may be at infinity. Even though poles p_i of $H(s)$ are transformed simply into poles $e^{p_i T_s}$ of $G(z)$, the transformation of the zeros is much more complicated. In the limiting case of small sampling time T_s, the location of the zeros of $G(z)$ are governed by the following theorem.

Theorem 4.1. (Åström *et al.* [4]). *Let $H(s)$ be the rational function*

$$H(s) = K \frac{(s - z_1)(s - z_2) \cdots (s - z_m)}{(s - p_1)(s - p_2) \cdots (s - p_n)} \tag{4.1}$$

with $m < n$ and $G(z)$ the corresponding pulse transfer function. Then, $T_s \to 0$ as, m zeros of $G(z)$ go to 1 as and the remaining $n - m - 1$ zeros of $G(z)$ go to zeros of the polynomial $B_{n-m}(z)$.

The polynomials $B_{n-m}(z)$ are given in [4,5] and have both minimum phase and non-minimum phase real zeros for $n - m \geq 2$. None of the NMP zeros of $B_{n-m}(z)$ are complex. So, $G(z)$ corresponding to $H(s)$ with relative degree greater than or equal to 2 ($n - m \geq 2$) will necessarily have real NMP zeros coming from $B_{n-m}(z)$ for small sampling time. Complex NMP zeros, if they exist, occur in complex conjugate pairs since $G(z)$ is real rational. The conditions under which complex NMP zeros occur for small sampling time can be determined by using Theorem 4.1 and are summarised by the following corollary.

Corollary 4.2. *For small T_s, the m_c ($2m_c < m$) complex conjugate zero pairs of $H(s)$ in (4.1), if any, result in m_c complex conjugate zero pairs of $G(z)$ which go to 1 as $e^{z_i T_z} = e^{T_s Re(z_i)} e^{j T_s Im(z_i)}$ as $T_s \to 0$.*

Some facts that follow from Theorem 4.1 and its corollary are listed below.

Fact 4.3. *For small T_s, complex zeros of $G(z)$ are only possible for $H(s)$ with $m \geq 2$.*

Fact 4.4. *For small T_s, $G(z)$ will have complex NMP zeros if and only if $H(s)$ has complex NMP zeros.*

The locations of zeros of $G(z)$ in the limiting case of large sampling time are characterised by the following theorem.

Theorem 4.5. (Åström *et al.* [4]). *Let $H(s)$ be a strictly proper, stable and rational transfer function with $H(0) \neq 0$. Then all zeros of $G(z)$ go to 0 as $T_s \to \infty$.*

Note that for large sampling time, complex zeros of $H(s)$ may become real zeros of $G(z)$. Also, it is not possible for $G(z)$ to have NMP zeros for large enough sampling time if $H(s)$ satisfies the conditions of Theorem 4.5. A conservative condition to guarantee minimum phase I is given in [4]. Less conservative conditions can be found in [6,7].

The conditions under which NMP zeros can occur for large sampling time can be determined by using Theorem 4.5 and are summarised by the following corollary.

Corollary 4.6. *For large enough sampling time, $G(z)$ can have NMP zeros if and only if $H(s)$ satisfies one or more of the conditions: $H(s)$ unstable, biproper or $H(0) = 0$.*

Example 4.1

Consider the continuous-time transfer function

$$H(s) = \frac{(s^2 - 4s + 8)}{s(s^2 + 2s + 5)}, \qquad (4.2)$$

with two complex NMP zeros at $s_1, \bar{s}_1 = 2 \pm j2$. Figure 4.3 shows a plot of the zeros of $G(z)$ corresponding to $H(s)$ in (4.2) as T_s is varied. It is seen that $G(z)$ has a complex conjugate NMP zero pair for small T_s as predicted by the corollary to Theorem 4.1. For large T_s, these zeros go to 0 as predicted by Theorem 4.5.

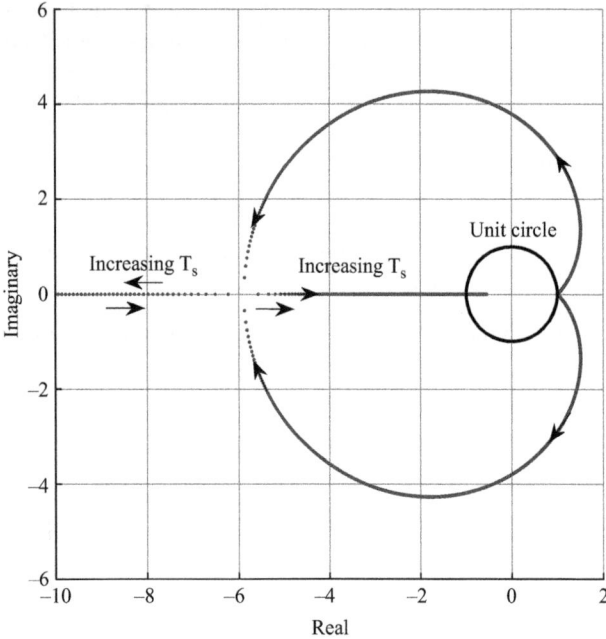

Figure 4.3   Location of zeros in Example 4.1 for Ts: $0 \to \infty$

Example 4.2 (Åström *et al.* [4])

Consider

$$H(s) = \frac{1}{(s+1)(s^2+1)},$$ (4.3)

which has no zeros but two unstable poles on the imaginary axis. So, one of the conditions of Theorem 4.5 is not satisfied and $G(z)$ can have NMP zeros for large enough sampling time by Corollary 4.6. $G(z)$ has complex NMP zeros for some choices of T_s as shown in Figure 4.4. Note also that the zeros of $G(z)$ do not go to zero as $T_s \to \infty$, since the conditions of Theorem 4.5 are not satisfied. In fact, the zeros start at $-2 \pm \sqrt{3}$ for $T_s = 0$, and traverse the complex plane periodically with a period in T_s of 2π; see Figure 4.4.

4.3   Different feedforward controller designs

There are several methods that can be used to design non-causal approximate inverse filters for discrete-time systems with NMP zeros. Some of the important methods are introduced in this section. The earliest method of designing preview-based reference

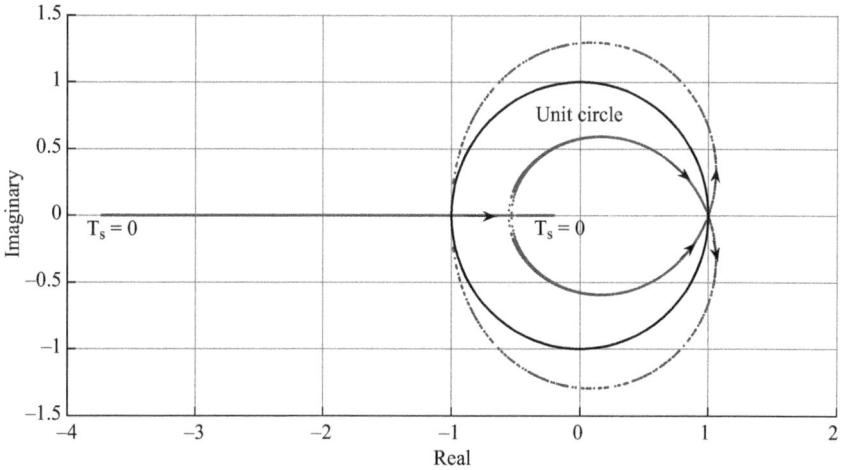

Figure 4.4   Location of zeros in Example 4.2 for Ts: 0 → ∞

feedforward controllers is the ZP compensation scheme of Tomizuka [8], where the phase error due to the NMP zeros is eliminated to achieve ZP error trajectory tracking. Gain errors due to the NMP zeros still exist and can be compensated for by using the E-filter [9], the ZPG method [10], the ZPGE method [11] or the ZPGO method [12], among other methods, where additional gain compensating zeros are used to reduce the gain error for NMP zeros while keeping ZP error characteristics. In the methods of ZPG and ZPGE, the gain compensating zero is determined using design equations based on a single point in the system frequency response to approximately obtain the desired bandwidth. The aim of ZPGO is to design the gain compensating zero optimally. In the TSA method of Gross *et al.* [13], polynomial long division is used to obtain a non-ZP solution. This solution can also be obtained via an optimisation approach.

4.4   Zero phase compensation

The transfer function of the nominal feedback control architecture in Figure 4.1 can be expressed as

$$G_{fb}(z) = \frac{n_{mp}(z)n_{nmp}(z)}{d(z)}, \qquad (4.4)$$

where $n_{mp}(z)$ and $n_{nmp}(z)$ are the minimum and non-minimum phase parts, respectively, of the numerator. The non-minimum phase part of the numerator can be factored into

$$n_{nmp}(z) = \prod_{k=1}^{m}(z - z_k), \qquad (4.5)$$

where $z_k \in \mathbb{C}$ such that $|z_k| \geq 1$. Thus, z_k's are the m non-minimum phase zeros of the feedback system. Noting that complex conjugation of factors of the form $z - z_k$ ($z_k \in \mathbb{C}$) over the unit circle amounts to

$$\overline{(z - z_k)}\big|_{z=e^{j\theta}} = (z^{-1} - \bar{z}_k)\big|_{z=e^{j\theta}}. \tag{4.6}$$

The ZP approximate inverse of $G_{fb}(z)$ is given by

$$\tilde{G}_{ZP}^{-1}(z) = \frac{d(z)}{n_{mp}(z)} \prod_{k=1}^{m} \frac{(z^{-1} - \bar{z}_k)}{(1 - z_k)(1 - \bar{z}_k)}. \tag{4.7}$$

The first part of (4.7), i.e., $d(z)/n_{mp}(z)$, is the invertible part of $G_{fb}(z)$. The remaining part of (4.7) has been formulated so that the compensated system $\tilde{G}_{ZP}^{-1}(z)G_{fb}(z)$ has ZP angle at all frequencies. Whatever can be inverted in $G_{fb}(z)$ has been inverted and the remaining non-minimum phase part has been multiplied by its complex conjugate in the frequency domain. The $(1 - z_k)$ type terms in the denominator of (4.7) are introduced to make the d.c. gain of the feedforward compensated system equal to unity. The transfer function of the ZP feedforward compensated system becomes

$$\tilde{G}_{ZP}^{-1}(z)\, G_{fb}(z) = \prod_{k=1}^{m} \frac{(z - z_k)(z^{-1} - \bar{z}_k)}{(1 - z_k)(1 - \bar{z}_k)}. \tag{4.8}$$

This method assures that the phase shift between the reference input and the output becomes zero for all frequencies. However, while the phase errors are cancelled, this approach may increase the gain errors. Therefore, desired tracking control characteristics can be obtained only within a relatively small bandwidth. One should therefore combine the ZP error of the ZP method with a gain error compensation method.

Assume that z_k's, which are the NMP zeros of the closed-loop system, can be factored into

$$n_{nmp}(z) = \prod_{k=1}^{m} (z - z_k), \tag{4.9}$$

where $z_k \in \mathbb{R}$ such that $|z_k| \geq 1$. Consider only one of these real NMP zero terms and adjust for unity d.c. gain as

$$\frac{z - z_k}{1 - z_k}. \tag{4.10}$$

Its frequency response is evaluated by letting $z = e^{j\theta}$ as

$$\frac{e^{j\theta} - z_k}{1 - z_k}. \tag{4.11}$$

Here, the angle θ with $0 \leq \theta \leq \pi$ is given by $\theta = \omega T_s$ where ω is the frequency and T_s is the sampling time. After manipulation, the frequency response magnitude and phase become

$$\left| \frac{z - z_k}{1 - z_k} \right|_{z=e^{j\theta}} = \sqrt{\frac{1 - 2z_k \cos \theta + z_k^2}{(1 - z_k)^2}}, \tag{4.12}$$

$$\angle \left(\frac{z - z_k}{1 - z_k} \right)_{z=e^{j\theta}} = \tan^{-1} \left(\frac{\sin \theta}{\cos \theta - z_k} \right). \tag{4.13}$$

By examining (4.12), the NMP zeros are further divided into two types as type 1 and type 2 zeros based on whether they lie to the left or to the right of the unit circle [10]. If $z_k < -1$, z_k is defined as a type 1 NMP zero and it is defined as a type 2 NMP zero if $z_k > 1$. The difference of these two types of NMP zeros can be seen from their amplitude Bode plots displayed in Figure 4.5. The first diagram in this figure shows that the gain of a type 1 NMP zeros attenuates from one as θ increases. On the other hand, the gain of a type 2 NMP zeros increases from one as θ increases. The presence of gain errors are evident from Figure 4.5. From the tracking viewpoint, these errors will severely degrade the tracking performance. The ZPG method of Menq and Xia [10] was formulated in order to compensate for the gain errors.

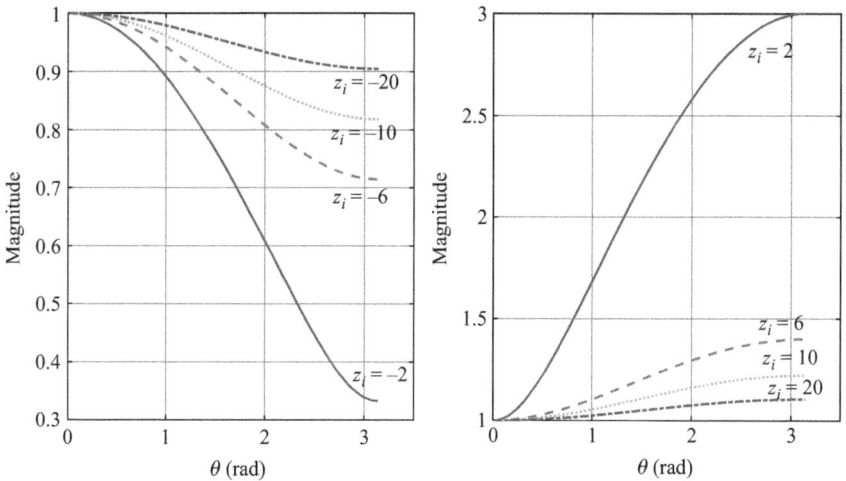

Figure 4.5   Normalised gains of Type 1 (left) and Type 2 (right) NMP zeros

4.5 Zero phase gain compensation

Gain errors due to type 1 NMP zeros can be compensated for by appropriately chosen type 2 NMP zeros in the feedforward compensator and vice versa. The ZP error property is preserved in this method. The ZPG feedforward compensator is given by

$$\tilde{G}_{ZPG}^{-1}(z) = \frac{d(z)}{n_{mp}(z)} \prod_{k=1}^{m} \frac{(z^{-1} - \bar{z}_k)}{(1 - z_k)(1 - \bar{z}_k)} \frac{(z - z_k^*)(z^{-1} - \bar{z}_k^*)}{(1 - z_k^*)(1 - \bar{z}_k^*)}, \tag{4.14}$$

where $z_k^* \in \mathbb{C}$ is the gain compensating zero corresponding to the NMP zero z_k. The ZPG feedforward compensator in (4.14) adds gain compensating zero expressions adjusted for ZP on top of the ZP controller of (4.7). The transfer function of the feedforward compensated system becomes

$$\tilde{G}_{ZPG}^{-1}(z)G_{fb}(z) = \prod_{k=1}^{m} \frac{(z - z_k)(z^{-1} - \bar{z}_k)}{(1 - z_k)(1 - \bar{z}_k)} \frac{(z - z_k^*)(z^{-1} - \bar{z}_k^*)}{(1 - z_k^*)(1 - \bar{z}_k^*)}, \tag{4.15}$$

and its frequency response is characterised by

$$\tilde{G}_{ZPG}^{-1}(z)G_{fb}(z) \bigg|_{z=e^{j\theta}} = \prod_{k=1}^{m} \frac{(e^{j\theta} - z_k)(e^{-j\theta} - \bar{z}_k)}{(1 - z_k)(1 - \bar{z}_k)} \frac{(e^{j\theta} - z_k^*)(e^{-j\theta} - \bar{z}_k^*)}{(1 - z_k^*)(1 - \bar{z}_k^*)}. \tag{4.16}$$

The gain compensating zeros are determined by using

$$\left| \tilde{G}_{ZPG}^{-1}(e^{j\theta})G_{fb}(e^{j\theta}) \right|_{\theta=\theta^*} = 1, \tag{4.17}$$

as the design equation. Here, θ^* is the desired bandwidth. In the case of a real NMP zero and a real gain compensating zero, the following equation is derived for the purpose of finding [10]:

$$\cos\theta^* = \frac{1}{2}\left[\frac{(1 - z_k)^2}{z_k} + \frac{(1 - z_k^*)^2}{z_k^*} \right] + 1. \tag{4.18}$$

Using (4.18), the desired bandwidth θ^* can be achieved approximately by selecting an appropriate z_k^*. From the design viewpoint, if the desired bandwidth θ^* is specified, the following equation can be obtained from (4.18) to determine the corresponding z_k^* that should be used in the feedforward compensator.

$$\frac{(1 - z_k^*)^2}{z_k^*} = -\left[\frac{(1 - z_k)^2}{z_k} + 2(1 - \cos\theta^*) \right]. \tag{4.19}$$

From the result of (4.19), a ZPG-based feedforward controller, which has ZP shift and small gain error up to the desired bandwidth, can be designed by substituting for z_k^* into (4.14).

By using ZPG, the system tracking bandwidth can be assigned arbitrarily according to the tracking signal and the tracking accuracy can be greatly improved. But it should be kept in mind that even though the ZPG design method usually results in satisfactory results, it is not an optimal method and there are situations where the desired bandwidth is not achieved. The ZPG method has been extended by Güvenç *et al.* [11] to cope with pairs of complex conjugate NMP zeros as well. A characterisation of complex non-minimum phase zeros is given in the next section.

4.6 Characterisation of complex non-minimum phase zeros

As mentioned before, Menq and Xia [10,14] have classified real NMP zeros as type 1 and type 2 zeros based on whether they lie to the left or to the right of the unit circle. Type 1 and type 2 NMP zeros have decreasing and increasing gain characteristics, respectively. Thus, the gain errors of type 1 zeros are compensated by introducing type 2 zeros and vice versa in the ZPG method of feedforward compensation. While such simple characterisation of complex NMP zeros is much more difficult because of the two coordinates involved rather than one, it is still useful to make some qualitative observations relating zero location in the complex plane to gain characteristics. For this purpose, the frequency response magnitude of the expression

$$[(z - z_1)(z - \bar{z}_1)] \left[\frac{1}{(1 - z_1)(1 - \bar{z}_1)} \right] \tag{4.20}$$

with complex NMP zero pair z_1 and $\bar{z}_1$ and where the second expression is for unity gain at zero frequency is considered. This is illustrated in Figure 4.6 where the magnitudes of $(z - z_1)$ and $(z - \bar{z}_1)$ are multiplied and recorded in a sweep of the point $z = e^{j\theta}$ with $\theta = \omega T_s$ from 0 to the Nyquist frequency of π radians to obtain the magnitude frequency response. Based on Figure 4.6, purely imaginary NMP zeros will

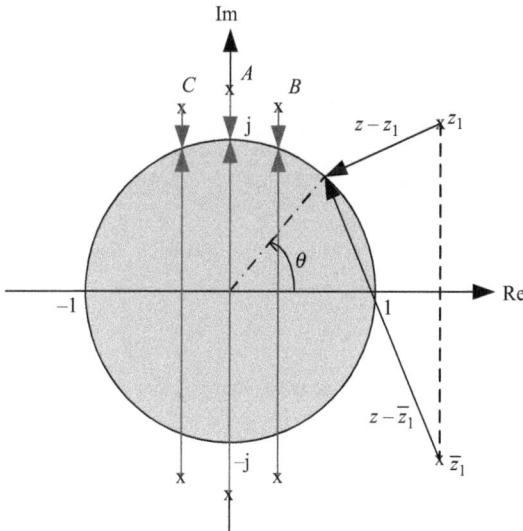

Figure 4.6   Illustration of discrete-time magnitude frequency response calculation for (4.20)

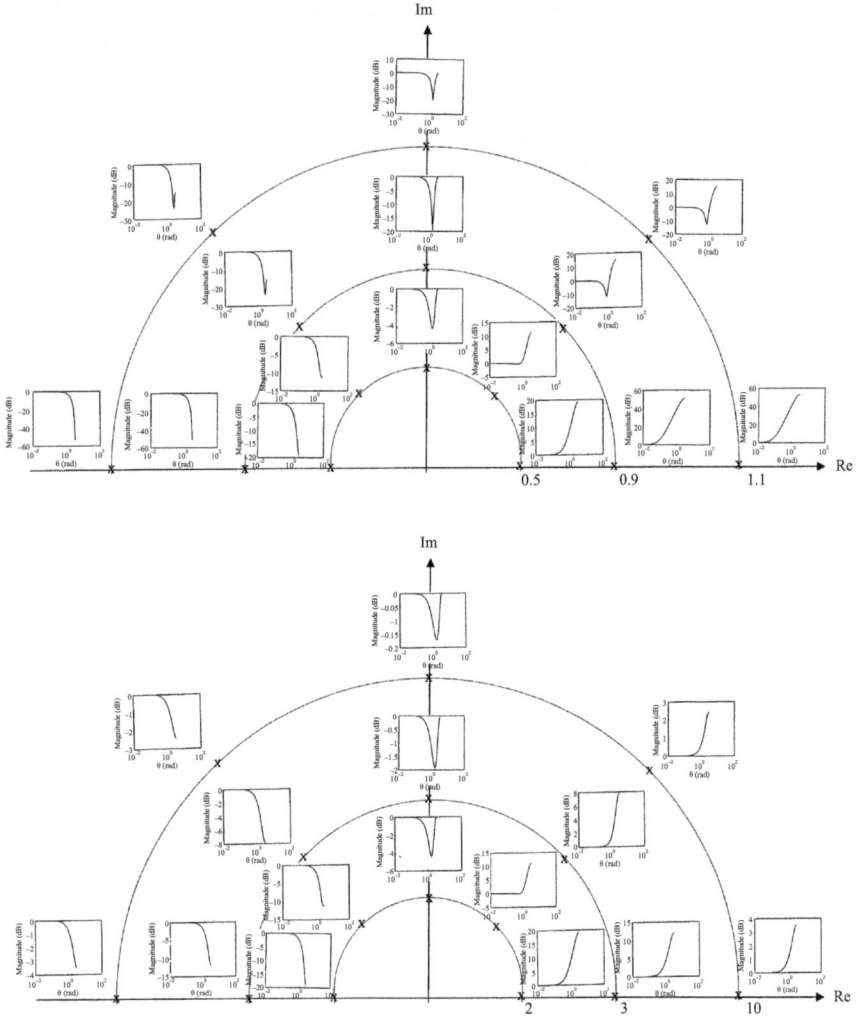

Figure 4.7   Effect of zero location on magnitude frequency response (4.20)
       (not to scale)

have a valley at $\theta = \pi/2$ in their magnitude frequency response that will get deeper as the zero approaches the unit circle (A in Figure 4.6). This phenomenon is observed in Figure 4.7 which shows the effect of NMP zero pair location on frequency response gain characteristics. Magnitude frequency responses of (4.20) at several different locations in the complex plane can be observed in Figure 4.7.

Analysis of Figures 4.6 and 4.7 reveals that for the right-half plane NMP zero pairs close to the imaginary axis, the valley occurs at $\theta < \pi/2$ (B in Figure 4.6) and for similar left-half plane NMP zero pairs the valley occurs at $\theta > \pi/2$ (C in Figure 4.6).

This effect becomes more pronounced as the NMP zero pair approaches the unit circle. Several observations follow.

Observation 4.7. *In principle, one can still categorise NMP zeros as type 1 and type 2 based on whether the NMP zero (pair, if complex) lie(s) in the left-half plane or right-half plane as these generally have decreasing and increasing gain characteristics. The valleys that occur for complex NMP zero pairs close to the imaginary axis and to the unit circle create problems in this line of reasoning. This is due to the corresponding valley in their magnitude frequency response plots that changes the general trend at intermediate frequencies.*

Observation 4.8. *Purely imaginary NMP zero pairs are the hardest to compensate for if the desired tracking bandwidth is $\theta^* > \pi/2$.*

Observation 4.9. *The gain errors of complex NMP zero pairs z_1, $\bar{z}_1$ sufficiently far away from the unit circle and the imaginary axis can be compensated for by introducing z_2 and $\bar{z}_2$ where $z_2 = -\text{Re}(z_1) - j\text{Im}(z_1)$ (opposite quadrant).*

Observation 4.10. *It is harder to compensate for gain errors of complex NMP zero pairs in comparison to real ones, this effect being more pronounced when the zeros are closer to the imaginary axis and the unit circle. Feedforward compensators therefore require more gain compensation zeros for systems with complex NMP zero pairs as compared to systems with real NMP zeros for comparable performance.*

It is clear from the qualitative characterisation above that compensating for gain errors is more involved for complex NMP zeros than it is for real NMP zeros. The ZPGE method which is an extension of the ZPG method to systems with complex NMP zero pairs is given in the following section. The optimal ZPG method of Aksun-Güvenç and Güvenç [12] given in Section 4.8 can also handle systems with complex NMP zeros in addition to real ones.

4.7 Zero phase extended gain compensation

In the ZPGE method, complex conjugate zero pairs can also be used as gain compensating zeros for real NMP zeros, to satisfy requirements like achieving a flat frequency response in addition to achievement of the desired bandwidth [11].

Consider the discrete-time transfer function

$$G(z) = \frac{n_{mp}(z)n_{nmp}(z)}{d(z)}, \tag{4.21}$$

where $n_{mp}(z)$ and $n_{nmp}(z)$ are the minimum phase and non-minimum phase parts, respectively, of the numerator. The non-minimum phase part of the numerator can be factored into

$$n_{nmp}(z) = \prod_{k=1}^{m_r}(z - z_k)\prod_{l=1}^{m_c}(z - z_l)(z - \bar{z}_l), \tag{4.22}$$

and has m_r real and m_c complex conjugate pairs of NMP zeros. Thus, $n_{nmp}(z)$ in (4.21) has a total of $m_r + 2m_c$ NMP zeros. The ZPGE approximate inverse filter for $G(z)$ in (4.21) is given by

$$\tilde{G}^{-1}_{\text{ZPGE}}(z) = \frac{d(z)}{n_{mp}(z)} \prod_{k=1}^{m_r} \frac{(z^{-1} - z_k)}{(1 - z_k)^2} \frac{(z - z^*_{k1})(z^{-1} - z^*_{k1})}{(1 - z^*_{k1})^2} \frac{(z - z^*_{k2})(z^{-1} - z^*_{k2})}{(1 - z^*_{k2})^2} .$$

$$\times \prod_{l=1}^{m_c} \frac{(z^{-1} - \bar{z}_l)(z^{-1} - z_l)}{(1 - \bar{z}_l)^2 (1 - z_l)^2} \frac{(z - z^*_{l1})(z^{-1} - z^*_{l1})}{(1 - z^*_{l1})^2} \frac{(z - z^*_{l2})(z^{-1} - z^*_{l2})}{(1 - z^*_{l2})^2} .$$

$$(4.23)$$

Note that the ZP shift requirement is met by pairing up terms of the form $(z - z_l)$ with terms of the form $(z^{-1} - \bar{z}_l)$. The product of each such pair has ZP shift on the unit circle $(z = e^{j\theta})$. The first part of (4.23), i.e., $d(z)/n_{mp}(z)$, is the invertible part of $G(z)$ in (4.21). The quantities in (4.23) with asterisks are the gain compensating zeros. z^*_{k1} and z^*_{k2} can be either both real or form a complex conjugate pair. Similarly z^*_{l1}, z^*_{l2} are either both real or form a complex conjugate pair. Thus five different combinations of NMP zeros and corresponding gain compensating zeros in $\tilde{G}^{-1}_{\text{ZPGE}}(z)$ are possible. These combinations are:

(a)   One real gain compensating zero corresponding to each real NMP zero z_k of $G(z)$. ($z^*_{k1} \in \mathbb{R}$, $z^*_{k2} = 0$, same as ZPG).
(b)   One complex conjugate gain compensating zero pair corresponding to each real NMP zero z_k of $G(z)$. (z^*_{k1}, $z^*_{k2} \in \mathbb{C}, \notin \mathbb{R}$. $z^*_{k1} = \bar{z}^*_{k2}$).
(c)   Two real gain compensating zeros corresponding to each real NMP zero z_k of $G(z)$. (z^*_{k1}, $z^*_{k2} \in \mathbb{R}$).
(d)   One complex conjugate gain compensating zero pair for each complex conjugate NMP zero pair z_l, $\bar{z}_l$ of $G(z)$. (z^*_{l1}, $z^*_{l2} \in \mathbb{C}, \notin \mathbb{R}$. $z^*_{l1} = \bar{z}^*_{l2}$).
(e)   Two real gain compensating zeros corresponding to each complex conjugate NMP zero pair z_l, $\bar{z}_l$ of $G(z)$. (z^*_{l1}, $z^*_{l2} \in \mathbb{R}$).

Define

$$B(\theta, z_i, z_j) = \left. \frac{(z - z_i)(z^{-1} - \bar{z}_i)(z - z_j)(z^{-1} - \bar{z}_j)}{(1 - z_i)(1 - \bar{z}_i)(1 - z_j)(1 - \bar{z}_j)} \right|_{z = e^{j\theta}}, \qquad (4.24)$$

where z_i and z_j are either both real or a complex conjugate pair. The desired tracking bandwidth θ^* can be approximated by the gain crossover frequency. This amounts to solving

$$\left. |\tilde{G}^{-1}(z)G(z)| \right|_{z = e^{j\theta^*}} = \prod_{k=1}^{m_r} B(\theta^*, z_k, 0) B(\theta^*, z^*_{k1}, z^*_{k2})$$

$$\times \prod_{l=1}^{m_c} B(\theta^*, z_l, \bar{z}_l) B(\theta^*, z^*_{l1}, z^*_{l2}) = 1. \qquad (4.25)$$

Order the $m_r + 2m_c$ NMP zeros of $G(z)$ with the m_r real ones followed by the m_c complex conjugate pairs. Solving

$$B(\theta^*, z_k, 0) \times B(\theta^*, z_{k1}^*, z_{k2}^*) = 1, \quad k \in \{1, 2, \ldots, m_r\} \tag{4.26}$$

$$B(\theta^*, z_l, \bar{z}_l) \times B(\theta^*, z_{l1}^*, z_{l2}^*) = 1, \quad l \in \{m_r + 1, \ldots, m_r + m_c\} \tag{4.27}$$

for the gain compensating zeros will result in θ^* being the gain crossover frequency. Equations (4.26) and (4.27) can be expanded into

$$(B_i - 1)\gamma_i^2 + \left[B_i(4\cos^2\theta^* - 2\alpha_i\cos\theta^* - 2) + 2\alpha_i - 2\right]\gamma_i$$
$$+ \left[B_i(\alpha_i^2 - 2\alpha_i\cos\theta^* + 1) - \alpha_i^2 + 2\alpha_i - 1\right] = 0, \tag{4.28}$$

where

$$\alpha_i = \begin{cases} z_{k1}^* + z_{k2}^* & \text{Cases b, c} \\ z_{l1}^* + z_{l2}^* & \text{Cases d, e} \end{cases} \tag{4.29}$$

$$\gamma_i = \begin{cases} z_{k1}^* z_{k2}^* & \text{Cases b, c} \\ z_{l1}^* z_{l2}^* & \text{Cases d, e} \end{cases} \tag{4.30}$$

$$B_i = \begin{cases} B(\theta^*, z_k, 0) & \text{Cases b, c} \\ B(\theta^*, z_l, \bar{z}_l) & \text{Cases d, e} \end{cases} \tag{4.31}$$

Case (a) (ZPG) was treated in detail earlier and hence is not included in (4.29)–(4.31). Note that B_i, α_i and γ_i are real for all the cases considered. Two design objectives are set for the determination of the gain compensating zeros as achieving the desired tracking bandwidth θ^* and making sure that the magnitude frequency response is sufficiently flat within that bandwidth. The design procedure is given in [11].

The disadvantage of relying on only the 0 dB crossover point for the desired bandwidth inherent in the ZPG method is also present in the ZPGE method. Therefore, an optimal method named ZPGO that utilises the precision tracking feedforward compensator but where the gain compensating zeros are determined so as to minimise a performance index is formulated next.

4.8 Zero phase optimal gain compensation

In the ZPGO method developed by Aksun-Güvenç and Güvenç [12], the performance index to be minimised is chosen as

$$\min_{z_k^*} \int_0^{\theta_{\max}} \left|1 - \frac{(z - z_k)(z^{-1} - z_k)}{(1 - z_k)^2} \frac{(z - z_k^*)(z^{-1} - z_k^*)}{(1 - z_k^*)^2}\right|^2_{z=e^{j\theta}} d\theta, \tag{4.32}$$

for $k \in \{1, 2, \ldots, m\}$ to determine the optimal ZPG gain compensating zero z_k^* where the second expression within the absolute value sign is the transfer function $\tilde{G}^{-1}(z)G(z)$ of the feedforward compensated system whose magnitude frequency response should be unity in the ideal case and a low-pass filter of desired tracking bandwidth in practice. Here, $[0, \theta_{max}]$ with $\theta_{max} \leq \pi$ is the frequency interval where the optimisation is performed. The performance index is modified as

$$\min_{z_k^*} \int_0^\pi \left| W(e^{j\theta}) \right| \left| 1 - \frac{(z - z_k)(z^{-1} - z_k)}{(1 - z_k)^2} \frac{(z - z_k^*)(z^{-1} - z_k^*)}{(1 - z_k^*)^2} \right|^2 \Bigg|_{z = e^{j\theta}} d\theta, \quad (4.33)$$

for $k \in \{1, 2, \ldots, m\}$, when the compensated transfer function is desired to be a low-pass filter. Both problems are solved using available optimisation routines.

Example 4.3

To demonstrate its use, the ZPGO method is applied to the model of a hydraulic positioning system used in non-circular turning (see, e.g., [10]) whose transfer function is given by

$$G(z) = \frac{0.02127\,(z + 0.1239)(z + 1.3231)}{(z - 0.952)\left[(z + 0.0621)^2 + (0.1677)^2\right]} = \frac{n_{mp}(z)n_{nmp}(z)}{d(z)}$$

$$n_{mp}(z) = 0.02127\,(z + 0.1239) \tag{4.34}$$

$$n_{nmp}(z) = z + 1.3231$$

where there is a real NMP zero at $z_1 = -1.3231$.

The ZPGO optimal feedforward compensator for this system is determined by using (4.32) with $\theta_{max} = 0.589\,\text{rad}$ as

$$\tilde{G}_{ZPGO}^{-1}(z) = \frac{d(z)}{n_{mp}(z)} \frac{(z^{-1} + 1.3231)}{(1 + 1.3231)^2} \frac{(z - 5.6608)(z^{-1} - 5.6608)}{(1 - 5.6608)^2}, \tag{4.35}$$

where $z_1^* = 5.6608$ is the optimal gain compensating zero.

The ZP feedforward compensator for this system is determined to be

$$\tilde{G}_{ZP}^{-1}(z) = \frac{d(z)}{n_{mp}(z)} \frac{(z^{-1} + 1.3231)}{(1 + 1.3231)^2}. \tag{4.36}$$

A comparison of the magnitude frequency responses of the ZP and ZPGO compensated systems shown in Figure 4.8 demonstrates the larger tracking bandwidth achieved by the ZPGO method. Accordingly, simulation results also show that the ZPGO compensated system achieves much lower tracking errors.

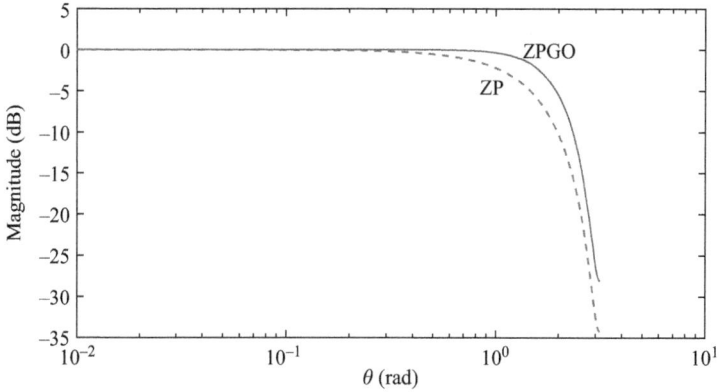

Figure 4.8   Comparison of gain frequency responses of ZP and ZPGO compensated systems

Even though, the method and the example considered used real NMP zeros, it will not be very difficult to extend the method to cope with pairs of complex conjugate NMP zeros. Consider the discrete-time transfer function given previously in (4.21).The ZPGO approximate inverse filter for this form of $G(z)$ is given by

$$\tilde{G}_{\text{ZPGO}}^{-1}(z) = \left[\frac{d(z)}{n_{mp}(z)}\right] \left[\prod_{k=1}^{m_{r1}} \frac{(z^{-1} - \bar{z}_k)}{(1 - z_k)^2} \prod_{l=1}^{m_{c1}} \frac{(z^{-1} - \bar{z}_l)(z^{-1} - z_l)}{(1 - \bar{z}_l)^2(1 - z_l)^2}\right]$$

$$\times \left[\prod_{r=1}^{m_{r2}} \frac{(z - z_r)(z^{-1} - z_r)}{(1 - z_r)^2}\right] \left[\prod_{c=1}^{m_{c2}} \frac{(z - z_c)(z^{-1} - \bar{z}_c)}{(1 - z_c)^2} \frac{(z - \bar{z}_c)(z^{-1} - z_c)}{(1 - \bar{z}_c)^2}\right], \quad (4.37)$$

where $z_r \in \mathbb{R}$ and $z_c \in \mathbb{C}$. The first part of (4.37), i.e., $d(z)/n_{mp}(z)$, is the invertible part of $G(z)$ as before. The second expression in (4.37) is for ZP of $n_{nmp}(z)$ after compensation and the last two expressions are for gain error compensation with m_{r2} real (cases (a), (c) and (e) of ZPGE) and m_{c2} complex conjugate pairs (cases (b) and (d) of ZPGE) of gain compensation zeros. The overall transfer function $\tilde{G}^{-1}(z)G(z)$, then, has, by construction, ZP at all frequencies and the gain compensation zeros are chosen optimally by solving

$$\min_{\substack{z_r \in \mathbb{R};\ r=1,2,\dots,m_{r2} \\ z_c \in \mathbb{C};\ c=1,2,\dots,m_{c2}}} \int_0^\pi \left|W(\theta)\left(1 - \tilde{G}^{-1}\left(e^{j\theta}\right) G\left(e^{j\theta}\right)\right)\right|^2 d\theta. \quad (4.38)$$

The filter $W(\theta)$ in (4.38) is a weighting function, which is usually chosen as

$$W(\theta) = \begin{cases} 1, & \theta \le \theta^*, \\ 0, & \theta < \theta^*, \end{cases} \quad (4.39)$$

where θ^* is the desired bandwidth of feedforward compensation. As an alternative, the weight can also be selected to minimise the tracking error for a given input [10], if the desired input trajectory is known in advance. The formulation in (4.39) works well for real as well as for most complex conjugate NMP zero pairs but there are some problems when the complex conjugate NMP zero pair is close to the imaginary axis and the unit circle. The formulation that works better in such cases is given in the following in the context of a simulation study on an example system.

Example 4.4

The closed-loop position control system of a hydraulic linear actuating system, used for tool positioning in non-circular machining by Tsao and Tomizuka [15] is chosen for the simulation study since its model has a complex conjugate NMP zero pair. The discrete-time, reduced order model of the closed-loop system is given by

$$G(z) = \frac{0.06z^2 + 0.034z + 0.071}{z^7 - 0.606z^6 - 0.747z^5 + 0.519z^4} = 0.06\frac{n_{nmp}(z)}{d_G(z)}. \qquad (4.40)$$

The complex NMP zeros are $z_1, \bar{z}_1 = -0.2833 \pm j1.0503$. This pair of complex conjugate NMP zeros is close to the imaginary axis and the unit circle and is thus hard to compensate for. The gain associated with it has a valley close to $\theta = \pi/2$rad and then rises to about unity at $\theta = \pi$ rad. In order to compensate for these gain characteristics, the compensating zeros need to introduce sufficient gain amplification at intermediate frequencies and sufficient gain attenuation at high frequencies. Since the optimisation formulation in (4.38) creates problems for these types of hard to compensate complex conjugate NMP zero pairs, it is replaced in this example by

$$\min_{\substack{z_r \in \mathbb{R}; r=1,2,...,m_{r2} \\ z_c \in \mathbb{C}; c=1,2,...,m_{c2}}} \int_0^\pi \left| 1 - W(\theta)\tilde{G}^{-1}\left(e^{j\theta}\right) G\left(e^{j\theta}\right) \right|^2 d\theta, \qquad (4.41)$$

where the weight $W(\theta)$ is chosen to have large gain at frequencies above θ^* and is chosen in this example as $3.61 / \left| e^{j\theta} + 0.9 \right|^2$ to provide large weight at high frequencies and, thus, force the compensated magnitude frequency response to a sharp decrease with frequency. The weight should be selected by balancing the desired tracking bandwidth requirement (uniform weight up to θ^*) with the low gain requirement at high frequencies (large weight above θ^*) for robustness of performance. The ZPGO computations, successively, with one complex conjugate gain compensation zero pair resulted in the best result being obtained by the two real zero case ($m_{r2} = 2$ and $m_{c2} = 0$ in (4.37)). The two real gain compensating zeros are $z_1 = 0.0867$ and $z_2 = 0.2874$. The magnitude frequency response plots in Figure 4.9 show the improvement over ZP compensation.

The response of the compensated system is simulated by using

$$r = 250 + 250\sin\left(\frac{2\pi}{0.05}t - \frac{\pi}{2}\right),$$ (4.42)

as the input. The sampling time used is 0.4 ms. The simulation results, depicted in Figure 4.10, show smaller tracking error by using the ZPGO method as compared to the ZP method.

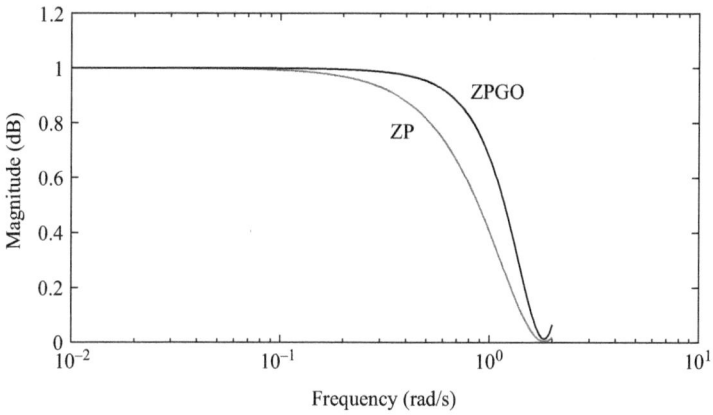

Figure 4.9   Magnitude frequency responses of ZPGO and ZP compensated systems

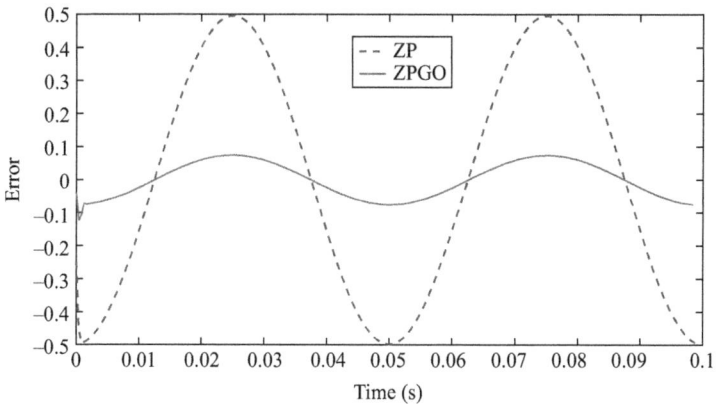

Figure 4.10   Simulated tracking errors

4.9   Truncated series approximation compensation

The TSA method of Gross *et al.* [13] which is the only non-ZP error algorithm considered here can be obtained either by polynomial long division or by solving the optimisation problem

$$
\min_{a_0,a_1,\ldots,a_p \in \mathbb{R}} \int_0^{2\pi} \left| \frac{1}{z - z_1} - (a_0 + a_1 z + a_2 z^2 + \cdots + a_p z^p) \right|^2_{z=e^{j\theta}} d\theta, \tag{4.43}
$$

where $\mathbb{R}$ is used to denote real numbers, z_1 is the NMP zero being compensated and the series expression in parentheses is the part of the feedforward compensator to be cascaded with the invertible part of the plant in forming the feedforward controller. Similar to the other methods, the TSA method can make the tracking degradation effect of NMP zeros arbitrarily small when more compensation terms are used. The method may be an alternative to ZP when a system, having NMP zeros, requires better gain characteristics, but it cannot be applied to zeros occurring on the unit circle.

   The number of required preview steps is dependent on the distance the NMP zero is from the unit circle.

4.10   Case study: electrohydraulic testing system

The example system considered here is an electrohydraulic material testing machine model given in [15]. The nominal model of this system under proportional feedback control K_p is

$$
C(s)P(s) = K_p \frac{K}{s\,(\tau_{sv}s + 1)\left(\frac{s^2}{\omega_h^2} + \frac{2\zeta_h s}{\omega_h} + 1\right)}, \tag{4.44}
$$

where $K_p K = 150$, $\tau_{sv} = 0.00187$ s, $\zeta_h = 0.4$ and $\omega_h = 618$ rad/s are the plant parameter values. The corresponding feedback loop with controller times plant given by (4.44) is discretised using the zero-order-hold method with a sampling time of 0.001 s. This results in a non-minimum phase zero at $z_1 = -7.9488$. The corresponding ZP, ZPG, ZPGO and TSA approximate feedforward inverse filters for this system are

$$
\tilde{G}_{\mathrm{ZP}}^{-1}(z) = \frac{d(z)}{n_{mp}(z)} \frac{0.0993z + 0.0125}{z}, \tag{4.45}
$$

$$
\tilde{G}_{\mathrm{ZPG}}^{-1}(z) = \frac{d(z)}{n_{mp}(z)} \frac{-0.0102z^3 + 0.118z^2 + 0.00482z - 0.00129}{z^2}, \tag{4.46}
$$

$$
\tilde{G}_{\mathrm{ZPGO}}^{-1}(z) = \frac{d(z)}{n_{mp}(z)} \frac{-0.0101z^3 + 0.118z^2 + 0.00491z - 0.00127}{z^2}, \tag{4.47}
$$

$$
\tilde{G}_{\mathrm{TSA}}^{-1}(z) = \frac{d(z)}{n_{mp}(z)} \left(-0.000203z^3 + 0.00201z^2 - 0.0158z + 0.126\right), \tag{4.48}
$$

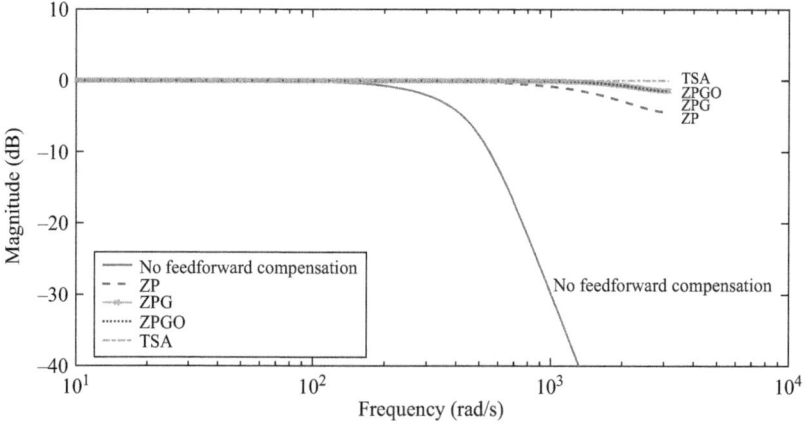

Figure 4.11   Magnitude frequency responses of all methods

where $\theta^* = \pi/5$ rad was specified as the desired bandwidth in the ZPG and ZPGO methods and a third-order truncated series was used in the TSA method. The approximate feedforward inverse filter compensated magnitude Bode plots for the designs in (4.45)–(4.48) are displayed in Figure 4.11.

4.11   Robustness analysis

Reference feedforward controllers cannot affect either the stability or the stability robustness of the loop being compensated. Robustness of performance of the feedforward compensated system is, thus, addressed here when the location of the zeros of $G(z)$ is uncertain. A robustness analysis similar to the one used by Gross *et al.* [13] is given here to address this issue. Only a complex NMP zero pair in the discrete-time plant is considered since the corresponding result for a real NMP zero is its special case. If a complex conjugate NMP zero pair, believed to be at z_n and $\bar{z}_n$ is actually at $z_n + \delta$ and $\bar{z}_n + \bar{\delta}$, respectively, the compensated transfer function becomes

$$\tilde{G}^{-1}(z)G(z) = \frac{\tilde{G}_n^{-1}(z)G_n(z)}{(z - z_n)(z - \bar{z}_n)} \left(z - (z_n + \delta)\right)\left(z - (\bar{z}_n + \bar{\delta})\right) \tag{4.49}$$

where $G_n(z)$ is the nominal (unperturbed) system and $\tilde{G}_n^{-1}(z)$ is the approximate inverse filter designed for it. Algebraic manipulations on (4.49) result in

$$\tilde{G}^{-1}(z)G(z) = \tilde{G}_n^{-1}(z)G_n(z) + \varepsilon(\delta, z) \tag{4.50}$$

where the additive term is

$$\varepsilon(\delta, z) = \frac{\tilde{G}_n^{-1}(z)G_n(z)}{(z - z_n)(z - \bar{z}_n)} \left(-\delta(z - \bar{z}_n) - \bar{\delta}(z - z_n) + \delta\bar{\delta}\right). \tag{4.51}$$

A conservative bound for this error is

$$|\varepsilon(\delta, z)| < \left| \frac{\tilde{G}_n^{-1}(z)G_n(z)}{(z - z_n)(z - \bar{z}_n)} \right| \left[(|z - \bar{z}_n| + |z - z_n|) |\delta| + |\delta|^2 \right]. \tag{4.52}$$

It is clear that as z_n comes closer to the unit circle and as the uncertainty δ decreases, the robustness of the magnitude frequency response will be better as a result of a smaller bound on $|\varepsilon(\delta, z)|$ in (4.52).

4.12   Chapter summary and concluding remarks

In this chapter, the effect of discrete-time real NMP zeros and complex conjugate NMP zero pairs to system performance were given in detail. The gain error associated with these NMP zeros was characterised based on their location in the complex plane. Notice that the location of zeros, being closer to the imaginary axis and the unit circle, lead to gain characteristics that are difficult to compensate for using preview-based feedforward control. For perfect tracking of a reference signal, we investigated several types of non-causal approximate inverse filters for discrete-time closed-loop systems:

- ZP (Zero Phase compensation)
- ZPG (Zero Phase Gain compensation)
- ZPGE (Zero Phase Extended Gain compensation)
- ZPGO (Zero Phase Optimal Gain compensation)

After comparing these four input shaping filters with each other, it was shown that the best result belongs to ZPGO when dealing with the system which contains NMP zeros. ZP can cancel the phase error but not cancel the all gain error. However, ZPG, ZPGE and especially ZPGO can eliminate both phase error and gain error.

References

[1]   B. Aksun-Güvenç, "Applied robust motion control," Ph.D. dissertation, Istanbul Technical University, Istanbul, Turkey, 2001.

[2]   K. Ogata, *Modern Control Engineering*.   Prentice Hall: New York, USA, 1990.

[3]   R. H. Cannon and E. Schmitz, "Initial experiments on the end-point control of a flexible one-link robot," *The International Journal of Robotics Research*, vol. 3, no. 3, pp. 62–75, Sept. 1984.

[4]   K. J. Åström, P. Hagander, and J. Sternby, "Zeros of sampled systems," *Automatica*, vol. 20, no. 1, pp. 31–38, 1984.

[5]   E. I. Jury, *Theory and Application of the z-Transform Method*.   Wiley, 1964.

[6]   Y. Fu and G. A. Dumont, "Choice of sampling to ensure minimum-phase behavior," *IEEE Transactions on Automatic Control*, vol. 34, no. 5, pp. 560–563, May 1989.

[7]   M. Ishitobi, "Conditions for stable zeros of sampled systems," *IEEE Transactions on Automatic Control*, vol. 37, no. 10, pp. 1558–1561, Oct. 1992.

[8]   M. Tomizuka, "Zero phase error tracking algorithm for digital control," *ASME Journal of Dynamic Systems, Measurement and Control*, vol. 109, no. 1, pp. 65–68, March 1987.

[9]   B. Haack and M. Tomizuka, "The effect of adding zeroes to feedforward controllers," *ASME Journal of Dynamic Systems, Measurement and Control*, vol. 113, no. 1, pp. 6–10, March 1991.

[10]   C.-H. Menq and Z. Xia, "Characterisation and compensation of discrete-time nonminimum-phase zeros for precision tracking control," in *Proceeding of the Symposium on Robotics Research*, 1990.

[11]   L. Güvenç, K. Harib, and K. Srinivasan, "Extended precision tracking control of discrete-time nonminimum phase systems," in *Proceedings of ASME International Mechanical Engineering Congress and Exposition*, 1995.

[12]   B. Aksun-Güvenç and L. Güvenç, "Optimal precision tracking control of discrete time nonminimum phase systems," in *Proceedings of European Control Conference*, 1999.

[13]   E. Gross, M. Tomizuka, and W. Messner, "Cancellation of discrete time unstable zeros by feedforward control," *ASME Journal of Dynamic Systems, Measurement and Control*, vol. 116, no. 1, pp. 33–38, March 1994.

[14]   J. Z. Xia and C.-H. Menq, "Precision tracking control of non-minimum phase systems with zero phase error," *International Journal of Control*, vol. 61, no. 4, pp. 791–807, 1995.

[15]   T.-C. Tsao and M. Tomizuka, "Robust adaptive and repetitive digital tracking control and application to a hydraulic servo for noncircular machining," *ASME Journal of Dynamic Systems, Measurement and Control*, vol. 116, no. 1, pp. 24–32, March 1994.

Chapter 5
Disturbance observer control

5.1 Introduction to disturbance observer control

The disturbance observer (model regulator is a particular method of designing a two-degree-of-freedom (2 DOF) control structure in order to achieve insensitivity against modelling errors due to model order reduction, linearisation and the presence of parameter uncertainties and also good sensor noise attenuation as well as disturbance rejection. It was introduced by Ohnishi [1] and further refined by Umeno and Hori [2]. It has been successfully applied in a variety of mechatronics applications. For instance, friction compensation in [3], in high-speed direct-drive positioning table [4], road vehicle yaw stability control in [5,6], robust AFM control in [7], power-assisted electric bicycle control in [8], table drive system in [9] and hard-disc-drive servo system in [10]. The augmentation of a plant with the disturbance observer forces it to behave just like its nominal (or desired) model within the bandwidth of disturbance observer. This is called model regulation. The disturbance observer also rejects disturbances within its bandwidth of operation and is, therefore, commonly used in disturbance rejection applications.

Unfortunately, the disturbance observer introduces stability and stability robustness problems due to the use of positive feedback. Classical stability robustness analysis for a disturbance observer control system uses unstructured multiplicative uncertainty and the Nyquist stability criterion. Many applications like those in References [11,12] involve well-defined real parametric uncertainty in a known plant. In such cases, a less conservative approach based on structured uncertainty like the SSV analysis can be used to investigate the effect of mixed parametric/complex uncertainty on stability and performance robustness of disturbance observer compensated systems [13]. The conventional approach to disturbance observer compensation is first introduced in this chapter. Later in the chapter, the parameter space approach is used for robustness analysis and design of disturbance observer controlled systems.

5.2 Continuous-time disturbance observer

In the disturbance observer approach, the inverse of the desired or nominal plant model is used to observe the disturbances and to cancel their effect through the control

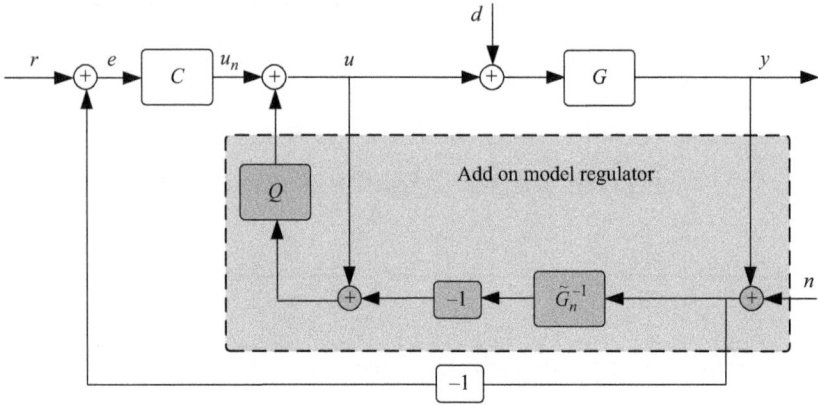

Figure 5.1   Feedback control system with add-on disturbance observer

signal. As a result, the closed system is forced to act like its nominal or desired model. The system structure with an add-on disturbance observer is shown in Figure 5.1.

Consider plant G with multiplicative uncertainty Δ_m and input disturbance d given by

$$y = G(u + d) \tag{5.1}$$

where $G = G_n(1 + \Delta_m)$ and G_n is the nominal model of the plant. The aim in the disturbance observer control is to obtain

$$y = G_n u_n \tag{5.2}$$

as the new input to output relation where u_n is the new control input. This aim can be achieved in the disturbance observer design by treating the external disturbance and model uncertainty as an extended disturbance e and solving for it as

$$y = G_n u + \underbrace{G_n d + G_n u \Delta_m + G_n d \Delta_m}_{e} \tag{5.3}$$

$$e = y - G_n u \tag{5.4}$$

and using the new control signal u_n given by

$$u = u_n - \frac{1}{G_n} e = u_n - \frac{1}{G_n} y + u \tag{5.5}$$

to approximately cancel its effect when substituted in (5.3). With the aim of not overcompensating at high frequencies and to avoiding stability robustness problems, the feedback signals in (5.5) are multiplied by the low-pass filter Q. In this case, (5.5) becomes

$$u = u_n - Q\left(\frac{1}{G_n}(y + n) + u\right) \tag{5.6}$$

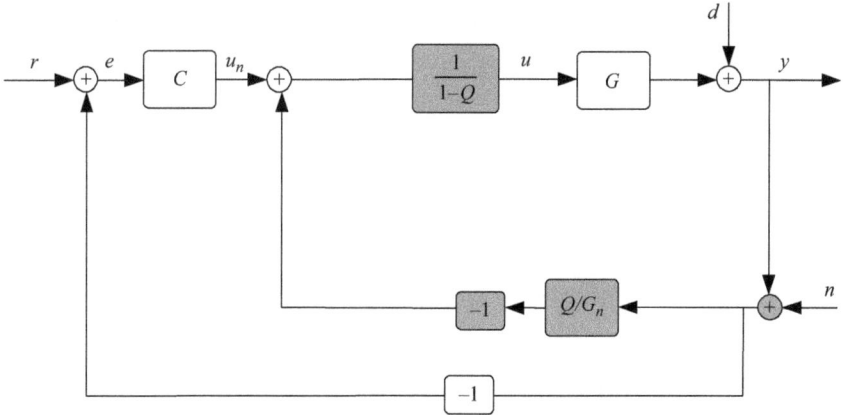

Figure 5.2   Feedback control system with modified disturbance observer

where n represents the sensor noise. Equation (5.6) can be used for a practical implementation of the disturbance observer and is illustrated in Figure 5.1.

An equivalent form of the disturbance observer is shown in the modified block diagram of Figure 5.2. Using Figure 5.2, the loop gain of the disturbance observer compensated plant without feedback control is

$$L = \frac{GQ}{G_n (1 - Q)} \tag{5.7}$$

with the model regulation, disturbance rejection and sensor noise rejection transfer functions given by

$$\frac{y}{u_n} = \frac{G_n G}{G_n (1 - Q) + GQ} \tag{5.8}$$

$$\frac{y}{d} = \frac{1}{1 + L} = \frac{G_n (1 - Q)}{G_n (1 - Q) + GQ} \tag{5.9}$$

$$\frac{y}{n} = \frac{-L}{1 + L} = \frac{-GQ}{G_n (1 - Q) + GQ} \tag{5.10}$$

It is seen that Q must be a unity gain low-pass filter. This choice will result in $y/u_n \rightarrow G_n$, $y/d \rightarrow 0$ at low frequencies where $Q \rightarrow 1$ and $y/n \rightarrow 0$ at high frequencies where $Q \rightarrow 0$. There are limitations in the selection of the bandwidth of the Q-filter. First of all, the bandwidth of the Q-filter should be allowed to exceed the bandwidth of the actuator used by at most a factor of two or three as it does not make sense to ask for control signals that the actuator cannot respond to. Another limitation for the Q-filter arises from the robust stability requirement based on unstructured plant model uncertainty.

The characteristic equation of the disturbance observer compensated system can be written as

$$G_n (1 - Q) + G_n (1 + \Delta_m) Q = 0 \qquad (5.11)$$

or

$$G_n (1 + \Delta_m) Q = 0 \rightarrow Q = -\frac{1}{\Delta_m} \qquad (5.12)$$

When the presence of Δ_m does not change the number of unstable poles and zeros of G in comparison to those of G_n, the application of the Nyquist stability criterion using (5.12) results in

$$|Q| < \left| \frac{1}{\Delta_m} \right|, \quad \forall \omega \qquad (5.13)$$

as the sufficient condition for robust stability. The conservative robust stability condition (5.13) is illustrated in Figure 5.3 for two different choices of the Q-filter, one that satisfies the condition and one that violates it.

The Q-filter design requirements outlined above are summarised graphically in Figure 5.2. The conventional approach to continuous-time disturbance observer compensator design is the choice of the two filters $Q(s)$ and $Q(s)/G_n(s)$ in Figure 5.1. Out of these, $G_n(s)$ is chosen first as the desired or nominal plant. $1/\Delta_m$ given by

$$\frac{1}{\Delta_m} = \frac{G_n}{G - G_n} \qquad (5.14)$$

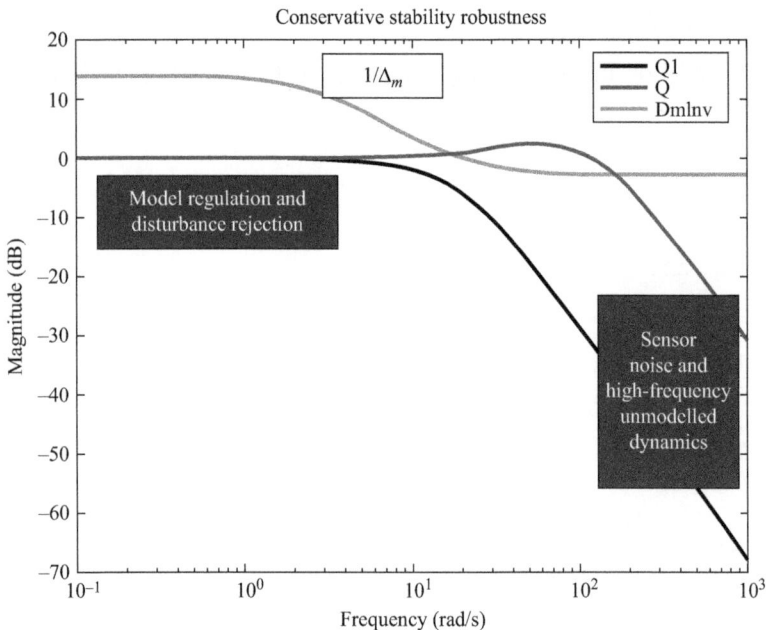

Figure 5.3   Summary of the Q-filter design requirements

forms a conservative upper bound for stability robustness on the magnitude of $Q(j\omega)$ according to (5.13). The filter $Q(s)$ is chosen as a low-pass filter with unity d.c. gain to satisfy the requirements in Figure 5.3. The order of $Q(s)$ is chosen to at least make the filter $Q(s)/G_n(s)$ causal and hence implementable. Some commonly used forms for $Q(s)$ are given in the equations below

$$Q(s) = \frac{1}{(\tau s + 1)^l} \tag{5.15}$$

$$Q(s) = \frac{1}{(s/\omega_n)^2 + (2\zeta/\omega_n)s + 1} \tag{5.16}$$

$$Q(s) = \frac{3\tau s + 1}{\tau^3 s^3 + 3\tau^2 s^2 + 3\tau s + 1} \tag{5.17}$$

All unity d.c. gain choices of $Q(s)$ in the disturbance observer compensation introduce integrators into the loop, resulting in reduced steady-state error for reference and disturbance inputs [13]. To see how the integrators are introduced, consider the $Q/(1-Q)$ part of the loop gain in (5.7) and consider $Q(s)$ of the form

$$Q(s) = \frac{\alpha_m s^m + \cdots + \alpha_{k+1} s^{k+1} + a_k s^k + a_{k-1} s^{k-1} + \cdots + a_1 s + 1}{a_n s^n + a_{n-1} s^{n-1} + \cdots + a_2 s^2 + a_1 s + 1}, \quad k \le m < n \tag{5.18}$$

$$\frac{Q}{1-Q} = \frac{\alpha_m s^m + \cdots + \alpha_{k+1} s^{k+1} + a_k s^k + a_{k-1} s^{k-1} + \cdots + a_1 s + 1}{s^{k+1} \left[a_n s^{n-k-1} + a_{n-1} s^{n-k-2} + \cdots + (a_m - \alpha_m) s^{m-k-1} \right.} \tag{5.19}$$
$$\left. + (a_{m-1} - \alpha_{m-1}) s^{m-k-2} + \cdots + (a_{k+1} - \alpha_{k+1}) \right]$$

which shows that $k+1$ integrators are incorporated into the loop for the choice of Q in (5.18). Note that all unity d.c. gain $Q(s)$ filters fit the form of (5.18) with $k \ge 1$ and hence add at least one integrator to the disturbance observer compensated system. For example, $Q(s)$ given by (5.15) and (5.16) add one integrator and $Q(s)$ given by (5.17) adds two integrators.

Let us now concentrate on the overall architecture with the feedback controller $C(s)$ in Figures 5.1 and 5.2. In this case, the closed-loop system input–output, disturbance rejection and sensor noise rejection transfer functions can be expressed as

$$\frac{y}{r} = \frac{CG_nG}{G_n(1-Q) + G(CG_n + Q)} \tag{5.20}$$

$$\frac{y}{d} = \frac{G_n(1-Q)}{G_n(1-Q) + G(CG_n + Q)} \tag{5.21}$$

$$\frac{y}{n} = \frac{-G(CG_n + Q)}{G_n(1-Q) + G(CG_n + Q)} \tag{5.22}$$

In the case of feedback control, the characteristic equation of the closed-loop system can be expressed as

$$G_n (1 - Q) + G_n (1 + \Delta_m)(CG_n + Q) = 0 \tag{5.23}$$

or

$$G_n (1 + CG_n + \Delta_m (CG_n + Q)) = 0 \rightarrow \frac{Q + CG_n}{1 + CG_n} = -\frac{1}{\Delta_m} \tag{5.24}$$

Application of the Nyquist stability criterion results in

$$\left| \frac{Q + CG_n}{1 + CG_n} \right| < \left| \frac{1}{\Delta_m} \right|, \quad \forall \omega \tag{5.25}$$

as the sufficient condition for robust stability including feedback control. Thus, the robust stability of the disturbance observer compensated system can be investigated in the absence and presence of the feedback control using (5.13) and (5.25), respectively.

5.3   Case study: Quanser QUBE™ Servo system

We will re-visit the Quanser QUBE™ servo angular position control problem that was considered in Section 3.6. The plant model from control input to motor angular position was

$$G_n (s) = \frac{28}{s (0.1s + 1)} = \frac{K_n}{s (\tau_n s + 1)}, \tag{5.26}$$

where the numerical values used were rounded to the nearest digit as $K_n = 28$ and $\tau_n = 0.1$s. This plant in (5.26) is chosen as the desired or nominal plant in the disturbance observer. Assume the presence of uncertainty in K_n and τ_n and a possible time delay τ_d that was neglected, resulting in the actual plant

$$G_n (s) = \frac{K}{s (\tau s + 1)} e^{-\tau_d s}. \tag{5.27}$$

The multiplicative uncertainty bound in (5.13) becomes

$$\frac{1}{\Delta_m(j\omega)} = \frac{G_n(j\omega)}{G(j\omega) - G_n(j\omega)} = \frac{K_n (\tau j\omega + 1)}{(\tau_n K e^{-\tau_d j\omega} - K_n \tau) j\omega + K e^{-\tau_d j\omega} - K_n}. \tag{5.28}$$

For this example, the time delay uncertainty τ_d is assumed to be zero. A 20% uncertainty in K_n and a 50% uncertainty in τ_n are assumed. The corresponding $1/\Delta_m$ uncertainty bound is shown in Figure 5.3.

Conventional disturbance observer design consists of choosing the filter G_n and the Q-filter which is chosen as the third-order low-pass unity d.c. gain filter with a relative degree of two given by

$$Q(s) = \frac{0.03s + 1}{(0.01)^3 s^3 + 3(0.01)^2 s^2 + 0.03s + 1} \tag{5.29}$$

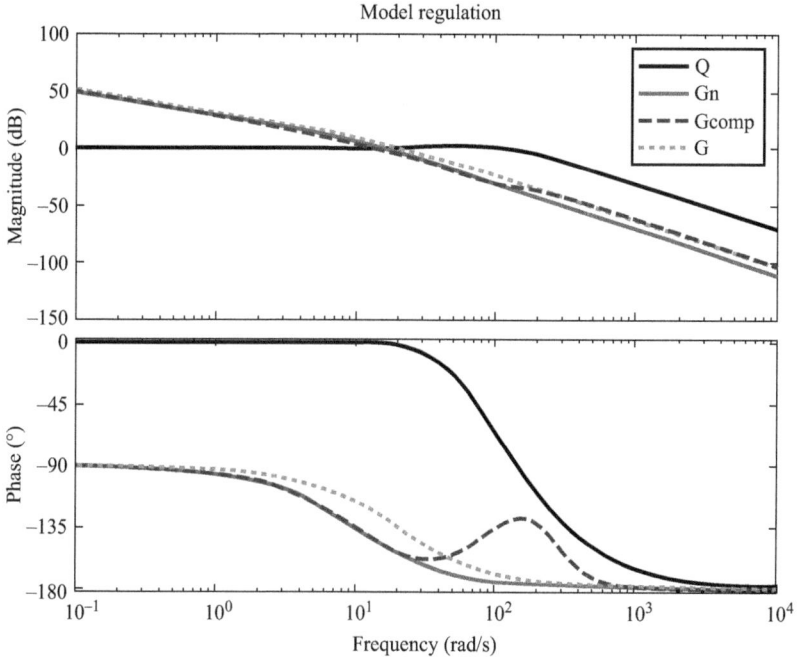

Figure 5.4   Model regulation of disturbance observer

$Q(s)$ is chosen as a second-order relative degree filter for causality of the other disturbance observer filter given by

$$\frac{Q(s)}{G_n(s)} = \frac{0.03s + 1}{(0.01)^3 s^3 + 3(0.01)^2 s^2 + 0.03s + 1} \frac{s(0.1s + 1)}{28} \tag{5.30}$$

The $Q(s)$ filter choice in (5.29) is shown as the dark gray line (Q) in Figure 5.3 which violates the robust stability condition (5.13). As (5.13) is a conservative condition, however, the resulting loop is still stable as will be seen from the following results.

The model regulation achieved by disturbance observer compensation as given by (5.8) is shown in Figure 5.4. An analysis of Figure 5.4 reveals that the large differences between the actual system (G) and its nominal model (G_n) is regulated to be close to G_n (G_{comp}) within the bandwidth of the disturbance observer Q. Magnitude model regulation is up to the bandwidth of the Q-filter whereas phase model regulation is effective up to one decade before this bandwidth. The disturbance rejection achieved by the disturbance observer is shown in Figure 5.5 where the disturbance rejection (DR) as given by (5.9)) is seen to be effective and low gain as desired up to the bandwidth of the Q-filter.

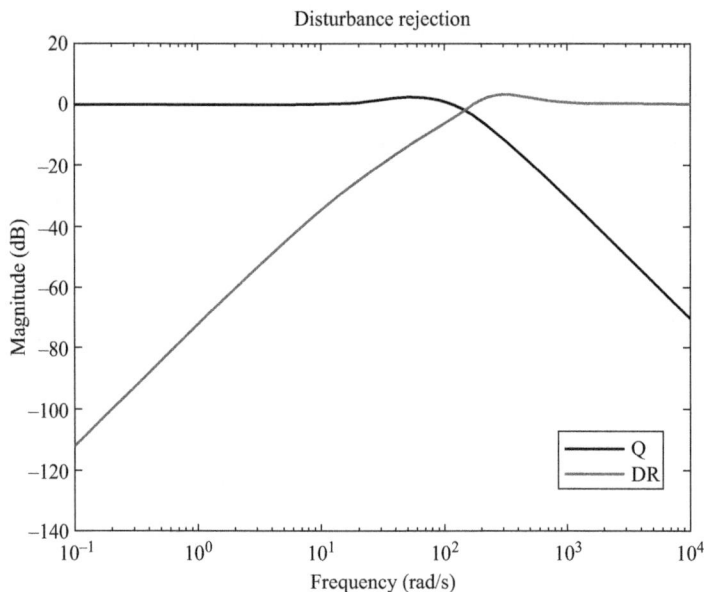

Figure 5.5   Disturbance rejection of disturbance observer

For this system a conventional feedback controller was already designed in Chapter 3 as

$$C_{\text{lead}}(s) = \frac{1.122s + 1}{9.238 \times 10^{-5}s + 1}.$$ (5.31)

Using this phase lead compensator and the disturbance observer design in (5.27) and (5.29) in the architecture of Figure 5.1 results in the model regulation frequency responses shown in Figure 5.6. These results show that the disturbance observer compensated closed-loop system characteristics (CL_{comp}) also follow that of the nominal system under feedback control (CL) up until the bandwidth of the Q-filter and follow that of the actual plant under feedback control (CL_{delta}) afterwards. If it is desired to obtain better model regulation at higher frequencies, the Q-filter bandwidth needs to be increased. The disturbance rejection of the feedback controlled system with ($DRCL_{comp}$) and without ($DRCL, DRCL_{delta}$) disturbance observer compensation are shown in Figure 5.7 which shows that the major strength of disturbance observer compensation is the exceptional disturbance rejection properties made possible by its use. The unit step disturbance input response in Figure 5.8 shows that the disturbance observer compensated system ($DRCL_{delta}$) has very good disturbance rejection properties and eliminates the step disturbance at steady state. The rate at which this disturbance is rejected can be adjusted by changing the bandwidth of the Q-filter.

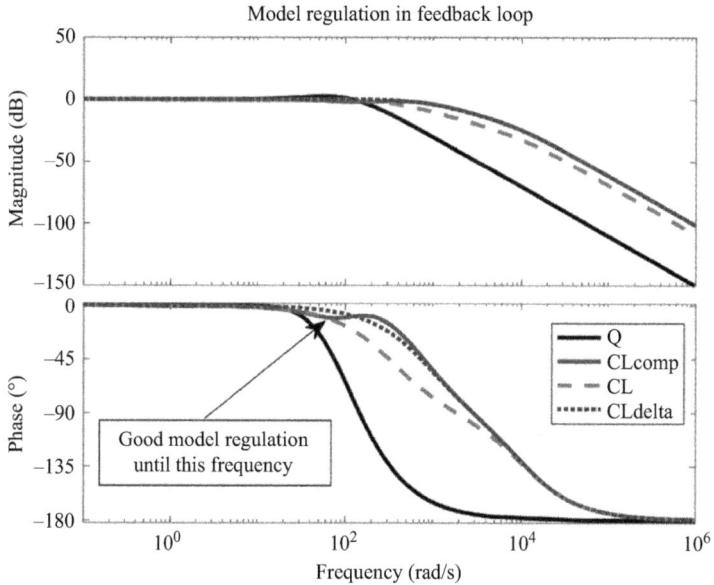

Figure 5.6 Model regulation of disturbance observer within feedback control loop

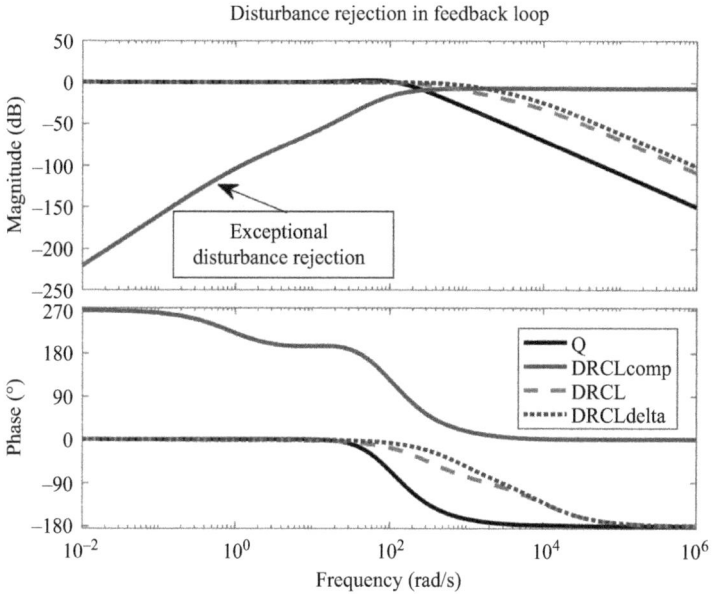

Figure 5.7 Disturbance rejection of disturbance observer within feedback control loop

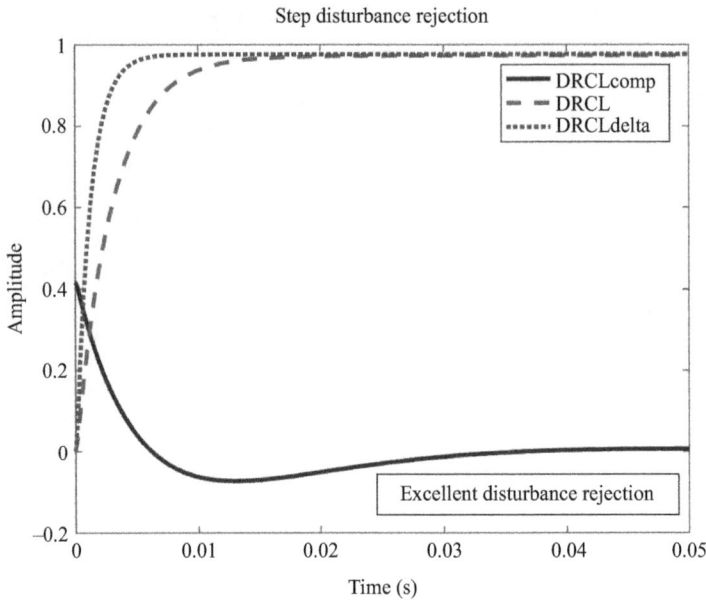

Figure 5.8   Unit step disturbance response

5.4   Mapping robust performance frequency domain specifications into disturbance observer parameter space

The conventional approach to disturbance observer design was presented in the previous sections. Parameter space methods are applied to disturbance observer design in this section. Similar to the parameter space approach given in [14] and also treated in Chapter 2 of this book, the parameter space design presented here is based on satisfying the mixed sensitivity requirement

$$\||W_S S| + |W_T T|\|_\infty < 1 \tag{5.32}$$

or equivalently satisfying

$$|W_S S| + |W_T T| < 1, \quad \forall \omega \in [0, \infty), \tag{5.33}$$

where S and T are the sensitivity and complementary sensitivity transfer functions and W_S and W_T are the corresponding weights. Under the limit condition, this requirement can be presented as

$$|W_S| + |W_T L| = |1 + L|, \quad \forall \omega \in [0, \infty), \tag{5.34}$$

which is called the point condition at each frequency. L in (5.34) is the loop gain. In order to obtain the region which satisfies expression (5.33) for all frequencies in the parameter space, (5.33) which is its boundary must be solved frequency at a time. The intersection of the curved lines for every calculated frequency leads to the

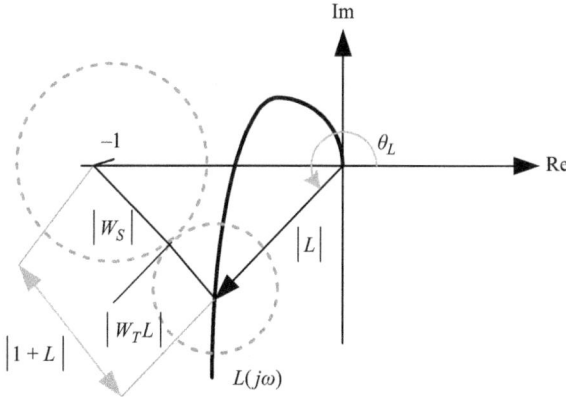

Figure 5.9 Point condition for the mixed sensitivity condition

overall boundary being searched for. The point condition is graphically illustrated in Figure 5.9.

By applying the cosine rule to the triangle in Figure 5.2, a graphical solution for $|L|$ results in

$$|L(j\omega)| = \frac{(-\cos\theta_L + |W_S(\omega)||W_T(\omega)|) \pm \sqrt{\Delta_L(\omega)}}{1 - |W_T(\omega)|^2} \tag{5.35}$$

where

$$\Delta_L = \cos^2\theta_L + |W_S|^2 + |W_T|^2 - 2|W_S||W_T|\cos\theta_L - 1 \tag{5.36}$$

When $L(j\omega)$, $W_S(j\omega)$ and $W_T(j\omega)$ are utilised in (5.35) and (5.36), the solutions for L at the chosen frequency ω are obtained. The solution procedure is to sweep angle θ_L from 0 to 2π radians and to solve for $|L|$ at each value of θ_L for which a solution exists. Then, all possible values of $L = |L|\,e^{j\theta_L}$ at the chosen frequency ω are obtained. Each value of L satisfies

$$L = KG = (K_R + jK_I)\,G \tag{5.37}$$

where K is the controller in an equivalent standard feedback architecture representation of the disturbance observer. The reader can be referred to Chapter 2 of this book and references [14,15] and the references therein for details of the general method of mapping frequency domain bounds to parameter space for standard feedback control architecture. For the case of a SISO system with disturbance observer compensation shown in Figure 5.1 (the inner loop without the feedback controller), the loop gain in (5.37) can be expressed as

$$L = KG = \left(\frac{Q}{G_n(1 - Q)}\right)G. \tag{5.38}$$

Then, the disturbance observer parameters within the filters G_n and Q can be solved for frequency at a time and mapped into the parameter space for robust design of SISO systems with disturbance observer compensation. For a specific disturbance observer filters choice given by

$$G_n = \frac{K_n}{\tau_n s + 1} \tag{5.39}$$

$$Q = \frac{1}{\tau_Q s + 1} \tag{5.40}$$

where K_n is the desired static gain of the disturbance observer controlled system, the solution can be obtained by solving for τ_Q and τ_n using

$$K_R + jK_I = \frac{L}{G} = \frac{Q}{G_n (1 - Q)} \tag{5.41}$$

$$K_R + jK_I = \frac{\tau_n j\omega + 1}{K_n (\tau_Q j\omega + 1)} \tag{5.42}$$

The solution procedure for τ_Q and τ_n leads to

$$\tau_Q = -\frac{1}{K_n K_I \omega} \tag{5.43}$$

$$\tau_n = -\frac{K_R}{K_I \omega} \tag{5.44}$$

as explicit solutions for this simple choice of the disturbance observer filters. Equations (5.43) and (5.44) result in straight lines in the $\tau_Q - \tau_n$ parameter space.

In order for more complicated and predefined structures of Q and G_n, a solution for τ_Q and τ_n exists and can also be found by either searching for a symbolic solution of the point condition equation or by solving it numerically. The final result is a region in a chosen controller parameter plane where the mixed sensitivity limit condition (5.33) is satisfied for the chosen frequency. The overall solution region, where the robust performance frequency domain bound is satisfied, can be obtained by means of repetition of the abovementioned procedure for a sweep of sufficiently many frequencies and the superposition of their results by graphical intersection in the chosen parameter space. The $\tau_Q - \tau_n$ controller parameter space is utilised for the selection of disturbance observer filters (5.39) and (5.40). The point condition (5.34) solution procedure is outlined below.

1. Choose a specific ω value. $|W_S (\omega)|$, $|W_T (\omega)|$ and $G (j\omega)$ at frequency ω are known at this point.
2. Let $\theta_L \in [0, 2\pi]$. Evaluate Δ_L by using (5.36) and select the active range of θ_L where $\Delta_L \geq 0$ is satisfied. For all values of θ_L in the active range:
    (a) Evaluate $|L|$ by using (5.35). Keep only the positive solutions.
    (b) Evaluate $L = |L| \, e^{j\theta_L}$.

(c) Solve for the corresponding disturbance observer controller real and imaginary parts K_R and K_I in (5.37).

(d) Substitute for K_R and K_I into right-hand sides of (5.43), and (5.44) to solve for τ_Q and τ_n.

3. Plot the closed curve obtained in the chosen controller parameter space. Either the inside or outside of this curve line is a solution of (5.33) at chosen frequency. The region obtained is the point condition solution in the chosen controller parameter space at the frequency chosen in Step 1.

4. Go back to Step 1 and repeat the procedure at a different frequency.

5. Plot the intersection of all point condition solutions for all frequencies considered. This is the overall solution region for the mixed sensitivity requirement.

The algorithm presented is a very fast numerical algorithm whose computation time depends on the number of grid points taken in the sweep of θ_L from 0 to 2π radians. Usually about 100 sweep points is taken for each frequency under consideration.

5.5 Case study: SISO disturbance observer control for yaw stability control of a road vehicle

The front wheel steering actuation based vehicle yaw stability control problem is utilised here as a numerical example in order to illustrate the design method developed in the previous section. The yaw stability controller in this example is designed using the linear single track vehicle model that was illustrated graphically in Figure 5.10. In this model, the transfer function from the front wheel steering angle δ_f input to the yaw rate r output is given by

$$G_{r\delta_f}(s, v) = \frac{r(s)}{\delta_f(s)} = \frac{b_1(v)s + b_0(v)}{a_2(v)s^2 + a_1(v)s + a_0(v)} \tag{5.45}$$

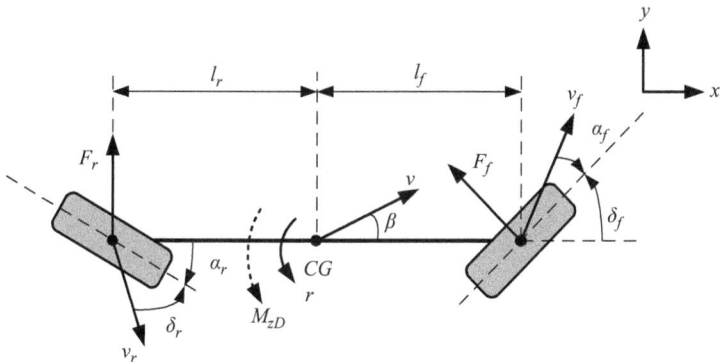

Figure 5.10   Single track vehicle model

with

$$b_0 = c_f c_r (l_f + l_r)v, \quad b_1 = c_f l_f m v^2$$
$$a_0 = c_f c_r (l_f + l_r)^2 + (c_r l_r - c_f l_f)m v^2 \tag{5.46}$$
$$a_1 = (c_f(I_z + l_f^2 m) + c_r(I_z + l_r^2 m))v, \quad a_2 = I_z m v^2$$

The design variables and their numerical values, which are used in the design and simulation, are given as follows: F_f (lateral force at front wheel), F_r (lateral force at rear wheel), M_{zd} (yaw disturbance moment), r (yaw rate), β (chassis side slip angle at vehicle centre of gravity), v (magnitude of vehicle velocity at centre of gravity), l_f (distance from front axle to centre of gravity, 1.25 m), l_r (distance from rear axle to centre of gravity, 1.32 m), δ_f (front wheel steering angle), δ_r (rear wheel steering angle set to zero in this section), m (vehicle mass, 1,296 kg), I_z (moment of inertia w.r.t. vertical axis at centre of gravity, 1,750 kg m^2), c_f (rear wheel cornering stiffness, 84,000 N/rad), c_r (rear wheel cornering stiffness, 96,000 N/rad), μ (road–tyre lateral friction coefficient). The numerical values presented corresponding to a mid-size passenger car.

The uncertainty region of friction coefficient versus vehicle speed is shown in Figure 5.11 where six exemplary points that will be used in the controller design are marked as R_1–R_6. The G_n and Q-filters are chosen to be of the form given by (5.39) and (5.40), where $K_n = K_n(v)$, the steady-state gain of the single track model at the chosen velocity, is used in (5.39). The inverse of the sensitivity function weight is chosen as

$$W_S^{-1}(s) = h_S \frac{s + \omega_S l_S}{s + \omega_S h_S} \tag{5.47}$$

with $l_S = 0.2$ (i.e., less than 20% steady-state error) being the low-frequency sensitivity bound, $h_S = 4$ being the high-frequency sensitivity bound and $\omega_S = 15$ rad/s being the approximate desired bandwidth of the disturbance observer compensator. The complementary sensitivity weighting function is chosen as

$$W_T(s) = h_T \frac{s + \omega_T l_T}{s + \omega_T h_T}, \tag{5.48}$$

where the low-frequency gain is $l_T = 0.5$, the high-frequency gain is $h_T = 1.5$ (corresponds to uncertainty of up to 150% at high frequencies) and the frequency of transition to significant model uncertainty is $\omega_T = 120$ rad/s. Check [15] for additional information on how to choose these weighting functions.

The solution procedure outlined in Section 5.4 results in the controller parameter space regions in Figure 5.12, each corresponding to one of the points R_1–R_6 in Figure 5.11. The controller parameters are chosen as $\tau_n = 0.15$ s and $\tau_Q = 0.02$ s and corresponds to a point within the solution regions for all six operating points. The corresponding yaw disturbance moment step input responses in Figure 5.13 exhibit very efficient and fast rejection of yaw disturbances in all cases.

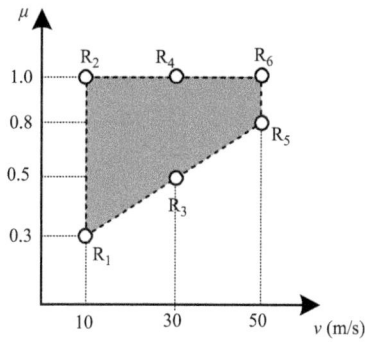

Figure 5.11   Uncertainty specifications

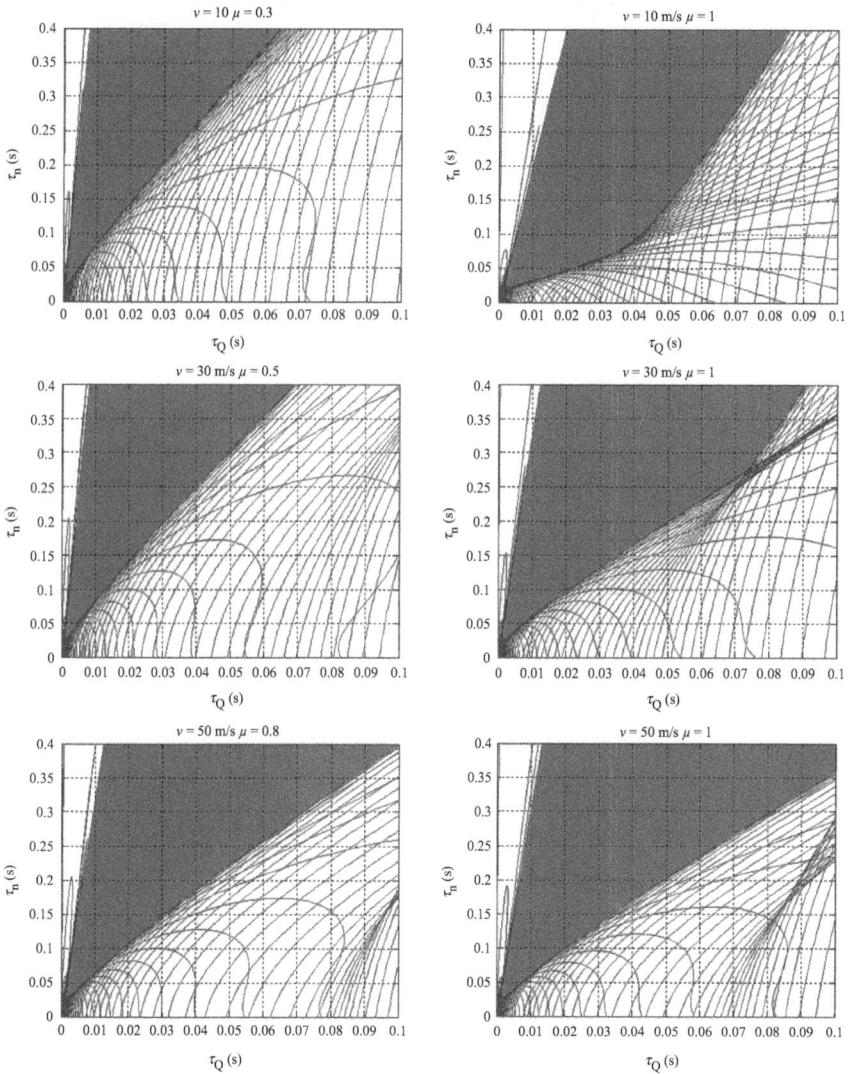

Figure 5.12   Solution regions in the parameter space for each of operating points

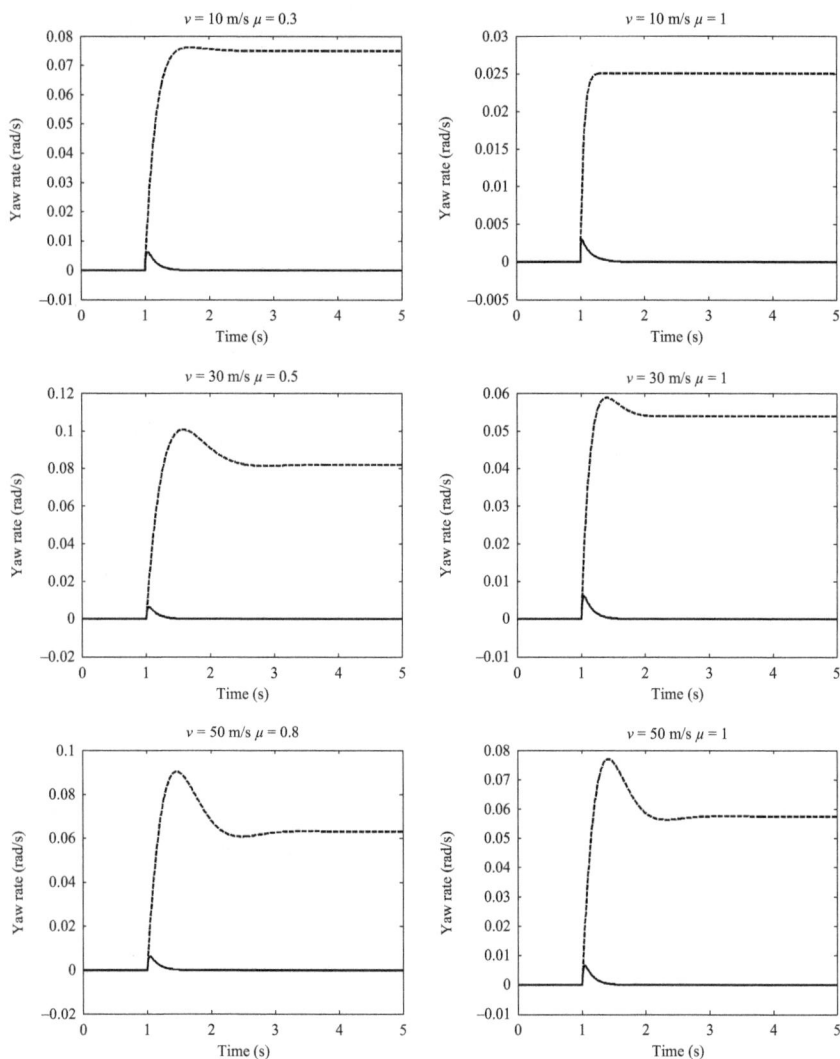

Figure 5.13   Simulation results for yaw moment disturbance step input (uncontrolled: dashed line, controlled: solid line)

5.6   Discrete-time disturbance observer

Discrete-time disturbance observer compensation was originally proposed for sensitivity reduction in systems with significant amounts of plant model uncertainty [16]. It has been applied successfully to cancelling the effect of non-linearities in a pneumatic system and friction compensation in motion and force control tasks [17,3].

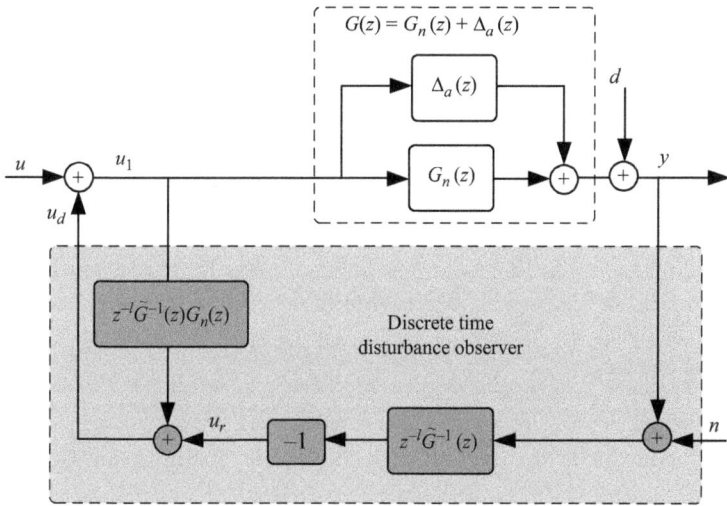

Figure 5.14 Discrete-time disturbance observer architecture

Most designers choose to use a continuous-time disturbance observer design procedure as it is possible to use high sampling rates and automatically convert their continuous-time design block diagram to a corresponding discrete-time controller that is implemented using rapid controller prototyping systems. While this approach is acceptable and results in good performance when high sampling rates are used, the designer, nevertheless, misses the extra design possibilities that are offered in a direct discrete-time design. A Q-filter is not needed for making the basic disturbance observer compensation implementable as difference equations are used.

The discrete-time disturbance observer architecture for a plant $G(z)$ including additive model uncertainty $\Delta_a(z)$ and subject to disturbance d is shown in Figure 5.14. $G_n(z)$ is the nominal or desired plant model and $\tilde{G}^{-1}(z)$ is the approximate inverse filter designed for it. The preview filter techniques introduced in Chapter 4 can be used for obtaining the approximate inverse filter $\tilde{G}^{-1}(z)$. Since the plant $G_n(z)$ can be non-minimum phase, an exact inverse is not used and is not recommended even for minimum phase plants for robustness reasons. Suitable approximate inverse filter design methods are the ZP, ZPG, ZPGE and ZPGO methods of Chapter 4. A delay of l steps, l being the relative degree of $\tilde{G}^{-1}(z)$, is used in the approximate inverse filter $z^{-l}\tilde{G}^{-1}(z)$ here as it has to be causal in contrast to available preview being used in Chapter 4. The filter $z^{-l}G_n(z)\tilde{G}^{-1}(z)$ in Figure 5.14 is like the Q-filter in the continuous-time implementation. While $G_n(z)\tilde{G}^{-1}(z)$ will require less than l steps of delay for causality, l steps of delay are, nevertheless, used in $z^{-l}G_n(z)\tilde{G}^{-1}(z)$ in order to make the delays of the two signals entering the disturbance observer comparison summer in Figure 5.14 similar.

For the standard discrete-time disturbance observer structure in Figure 5.14, block diagram algebra results in

$$Y(z) = \frac{G(z)}{1 - z^{-l}\tilde{G}^{-1}(z)\,(G_n(z) - G(z))}U(z) + \frac{1 - z^{-l}\tilde{G}^{-1}(z)G_n(z)}{1 - z^{-l}\tilde{G}^{-1}(z)\,(G_n(z) - G(z))}D(z)$$

$$- \frac{z^{-l}\tilde{G}^{-1}(z)G(z)}{1 - z^{-l}\tilde{G}^{-1}(z)\,(G_n(z) - G(z))}N(z) \tag{5.49}$$

Model regulation and disturbance rejection require the determination of $\tilde{G}^{-1}(z)$ to achieve $z^{-l}\tilde{G}^{-1}(z)G_n(z) \cong 1$. Stability of the discrete-time disturbance observer loop can be analysed by applying the Nyquist stability criterion [16] to the characteristic equation

$$1 - z^{-l}\tilde{G}^{-1}(z)\,(G_n(z) - G(z)) = 1 + z^{-l}\tilde{G}^{-1}(z)\Delta_a(z) = 0, \tag{5.50}$$

where the nominal plant $G_n(z)$ (with $\Delta_a \equiv 0$) is stable. Then, the condition

$$\left|\Delta_a(e^{j\omega T_s})\right| < \frac{1}{\left|\tilde{G}^{-1}\left(e^{j\omega T_s}\right)\right|} \quad \forall \omega \in [0, \infty), \tag{5.51}$$

derived from (5.50) is sufficient for stability robustness of the discrete-time disturbance observer loop if the phase of Δ_a is arbitrary and only a sufficient condition otherwise. Here, T_s in (5.51) is the sampling time.

5.7 MIMO decoupling extension of disturbance observer

The MIMO disturbance observer is developed for plants where the desired dynamics is decoupled. There is a wealth of practical applications where this is true and two case studies are presented later in this chapter. There are also important pitfalls of the approach presented here which are: (i) that a centralised MIMO decoupling disturbance observer is not presented, (ii) that the proposed method is not applicable to non-square MIMO systems and (iii) that the proposed method is not applicable to systems where decoupled loops (i.e., a diagonal MIMO transfer function matrix) are not desired.

Consider plant $\mathbf{G}$ with multiplicative model error $\mathbf{W_m \Delta_m}$ and external disturbance d. Its input–output relation can be expressed as

$$\mathbf{y} = \mathbf{Gu} + \mathbf{d} = (\mathbf{G_n(I + W_m \Delta_m)})\,\mathbf{u} + \mathbf{d} \tag{5.52}$$

where $\mathbf{G_n}$ is the MIMO nominal (or desired) model of the plant and $\mathbf{I}$ is the identity matrix of appropriate dimensions. $\mathbf{G}$ and $\mathbf{G_n}$ are square transfer function matrices with m inputs and m outputs and $\mathbf{G_n}$ is chosen to be non-singular. $\mathbf{G_n}$ is chosen as a diagonal matrix in a decoupling type design as

$$\mathbf{G_n}(s) = \mathbf{diag}\,\{G_{n1}(s), G_{n2}(s), \ldots, G_{nm}(s)\} \tag{5.53}$$

This approach treats MIMO disturbance observer compensation where the desired dynamics of the plant means a decoupled nominal or desired model G_n. Consequently,

(a)

(b)

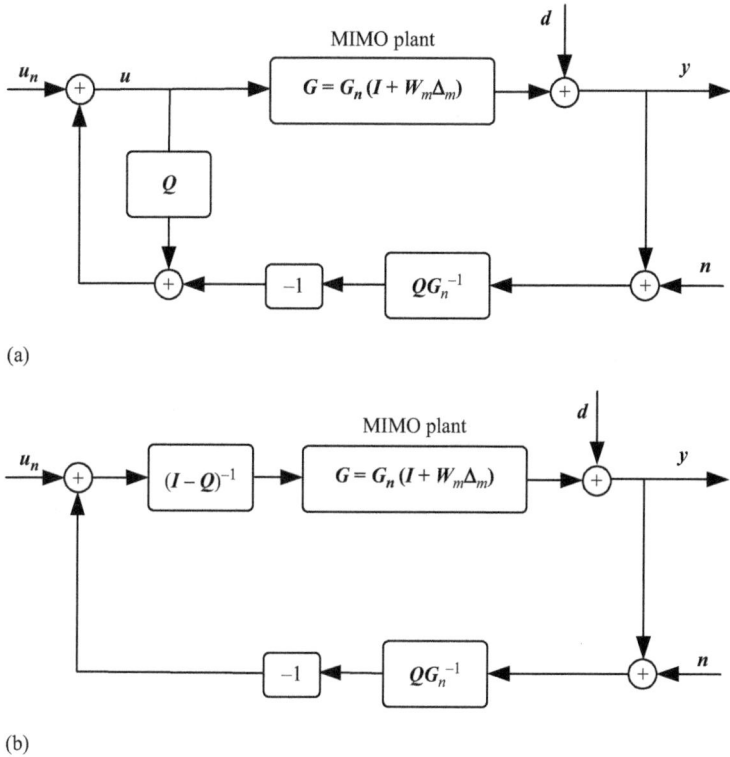

Figure 5.15 Plant with (a) disturbance observer compensation and (b) equivalent block diagram

the MIMO loop becomes m decoupled SISO loops which can be designed independently using the methods presented in the previous sections. The aim in decoupling type MIMO disturbance observer design is to obtain

$$\mathbf{y} = \mathbf{G_n u_n} \tag{5.54}$$

as the input–output relation in the presence of model uncertainty and external disturbance. $\mathbf{u_n}$ in (5.54) is the new input signal in Figure 5.15.

In decoupling MIMO disturbance observer design, the model regulation objective (5.54) is achieved by treating the external disturbance and model uncertainty as an extended disturbance $\mathbf{e}$ and solving for it as

$$\mathbf{y} = \mathbf{G_n u} + (\mathbf{G_n W_m \Delta_m u} + \mathbf{d}) = \mathbf{G_n u} + \mathbf{e} \tag{5.55}$$

$$\mathbf{e} = \mathbf{y} - \mathbf{G_n u} \tag{5.56}$$

and introducing

$$\mathbf{u} = \mathbf{u_n} - \mathbf{G_n^{-1} e} = \mathbf{u_n} - \mathbf{G_n^{-1} y} + \mathbf{u} \tag{5.57}$$

to cancel the effect of the extended disturbance $\mathbf{e}$ in (5.55). The extended disturbance $\mathbf{e} = \mathbf{G_n W_m \Delta_m u} + \mathbf{d}$ in (5.55) and (5.56) includes the effects of both plant modelling error including deviation from the desired decoupled dynamics represented by $\mathbf{\Delta_m}$ and disturbances represented by d. $\mathbf{G_n^{-1}}$ is given by

$$\mathbf{G_n^{-1}}(s) = \mathbf{diag}\left\{\frac{1}{G_{n1}(s)}, \frac{1}{G_{n2}(s)}, \dots, \frac{1}{G_{nm}(s)}\right\} = \mathbf{diag}\left\{\frac{1}{G_{ni}(s)}\right\} \qquad (5.58)$$

and has possibly non-causal transfer functions $1/G_{ni}(s) : i = 1, \dots, m$ for strictly proper $G_{ni}(s)$. Since this is usually not the case, $\mathbf{G_n^{-1}}$ is multiplied by the diagonal matrix of unity gain low-pass filters

$$\mathbf{Q}(s) = \mathbf{diag}\{Q_1(s), Q_2(s), \dots, Q_m(s)\} = \mathbf{diag}\{Q_i(s)\} \qquad (5.59)$$

such that $\mathbf{QGn^{-1}}$ and hence $Q_i(s)/G_{ni}(s) : i = 1, \dots, m$ are all causal. Another use of the $Q_i(s)$ filters is to limit the compensation to a low-frequency range to avoid saturating actuators and to avoid stability robustness problems due to high-frequency unmodelled dynamics. The feedback signals in the control law (5.57) are pre-multiplied by $\mathbf{Q}$, resulting in

$$\mathbf{u} = \mathbf{u_n} - \mathbf{QG_n^{-1}}(\mathbf{y} + \mathbf{n}) + \mathbf{Qu} \qquad (5.60)$$

where $\mathbf{y} + \mathbf{n}$ with $\mathbf{n}$ representing the sensor noise is used instead of $\mathbf{y}$. The MIMO disturbance observer compensation defined by (5.60) is illustrated in Figure 5.15. The use of (5.58) and (5.59) for the MIMO disturbance observer compensation filters $\mathbf{Q}$ and $\mathbf{G_n^{-1}}$ results in m SISO disturbance observer loops. The interactions between these loops in the original plant $\mathbf{G}$ are seen as disturbances and are rejected by the decoupling MIMO disturbance observer presented here.

The equation relating the output $\mathbf{y}$ in Figure 5.15 to the command input $\mathbf{u_n}$, disturbance input $\mathbf{d}$ and the sensor noise input $\mathbf{n}$ is

$$\mathbf{y} = \left[\mathbf{I} + \mathbf{G}(\mathbf{I} - \mathbf{Q})^{-1}\mathbf{QG_n^{-1}}\right]^{-1}\left\{\mathbf{G}(\mathbf{I} - \mathbf{Q})^{-1}\mathbf{u_n} + \mathbf{d} - \mathbf{G}(\mathbf{I} - \mathbf{Q})^{-1}\mathbf{QG_n^{-1}n}\right\} \quad (5.61)$$

The decoupling MIMO disturbance observer has the two major functions of disturbance and coupled response rejection and decoupled model regulation. Adequate sensor noise rejection is also desirable but neither the SISO nor the MIMO disturbance observer compensators will be able to help with sensor noise rejection. Consider the disturbance rejection transfer function matrix from disturbance input $\mathbf{d}$ to output $\mathbf{y}$ in (5.61) which can be converted into

$$\left[\mathbf{I} + \mathbf{G}(\mathbf{I} - \mathbf{Q})^{-1}\mathbf{QG_n^{-1}}\right]^{-1} = \mathbf{I} - \mathbf{G}(\mathbf{QG_n^{-1}G} + \mathbf{I} - \mathbf{Q})^{-1}\mathbf{QG_n^{-1}} \qquad (5.62)$$

using the matrix inversion lemma (see [18], for the matrix inversion lemma). Assuming $\mathbf{Q}(j\omega) = \mathbf{I}$ at low frequencies, the expression in (5.62) will become the zero matrix $\mathbf{0}$ and the disturbance rejection goal will be achieved. Consider the model regulation transfer function matrix from command input $\mathbf{u_n}$ to output $\mathbf{y}$ in (5.61). The use of the matrix inversion lemma followed by some manipulations results in

$$\left[\mathbf{I} + \mathbf{G}(\mathbf{I} - \mathbf{Q})^{-1}\mathbf{QG_n^{-1}}\right]^{-1}\mathbf{G}(\mathbf{I} - \mathbf{Q})^{-1} = \left\{(\mathbf{I} - \mathbf{Q})\mathbf{G^{-1}} + \mathbf{QG_n^{-1}}\right\}^{-1} \qquad (5.63)$$

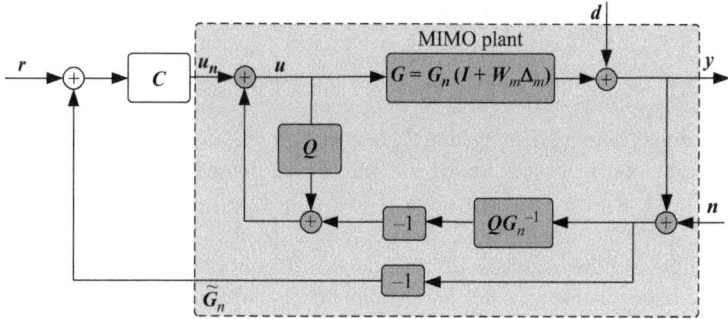

Figure 5.16 Plant under disturbance observer compensation and its feedback control

which becomes $\mathbf{G_n}$ as is desired when $\mathbf{Q}(j\omega) = \mathbf{I}$ is selected at low frequencies where decoupling MIMO model regulation is desired. The model regulation goal is achieved for $\mathbf{Q}(j\omega) = \mathbf{I}$, ensuring that the decoupled desired dynamics in $\mathbf{G_n}$ and robustness to plant modelling error $\mathbf{\Delta_m}$ and disturbances $\mathbf{d}$ within the extended disturbance $\mathbf{e}$ are achieved. Consider the sensor noise rejection transfer function matrix

$$-\left[\mathbf{I} + \mathbf{G}(\mathbf{I} - \mathbf{Q})^{-1}\mathbf{Q}\mathbf{G_n^{-1}}\right]^{-1}\mathbf{G}(\mathbf{I} - \mathbf{Q})^{-1}\mathbf{Q}\mathbf{G_n^{-1}} \tag{5.64}$$

from sensor noise input n to output y. We desire this transfer function matrix to ideally be the zero matrix 0 at high frequencies. This disturbance rejection goal is achieved if $\mathbf{Q}(j\omega) = \mathbf{0}$ at high frequencies where sensor noise occurs, meaning that the disturbance observer is shut off at those frequencies and will not be able to adversely affect sensor noise propagation within the loop. Ideal disturbance observer operation requires $\mathbf{Q}(j\omega) = \mathbf{I}$ at low frequencies (model regulation and disturbance rejection) and $\mathbf{Q}(j\omega) = \mathbf{0}$ at high frequencies (sensor noise rejection). The choice of $\mathbf{Q}$ as a diagonal matrix of unity d.c. gain low-pass filters as given in (5.59) satisfies these requirements at both low and high frequencies.

With the addition of the feedback controller $\mathbf{C}$ to the disturbance observer compensated system in Figure 5.15, the block diagram in Figure 5.16 is obtained. The equation relating the reference input $\mathbf{r}$, the disturbance $\mathbf{d}$ and the sensor noise $\mathbf{n}$ to the output $\mathbf{y}$ becomes

$$\mathbf{y} = \left\{\mathbf{I} + \mathbf{G}(\mathbf{I} - \mathbf{Q})^{-1}\left(\mathbf{C} + \mathbf{Q}\mathbf{G_n^{-1}}\right)\right\}^{-1}$$
$$\times \left\{\mathbf{G}(\mathbf{I} - \mathbf{Q})^{-1}\mathbf{C}\mathbf{r} - \mathbf{G}(\mathbf{I} - \mathbf{Q})^{-1}\left(\mathbf{C} + \mathbf{Q}\mathbf{G_n^{-1}}\right)\mathbf{n} + \mathbf{d}\right\} \tag{5.65}$$

The use of $\mathbf{Q}(j\omega) = \mathbf{I}$ at low frequencies results in ideal disturbance rejection; and reference command following corresponding to the disturbance observer augmented system in Figure 5.16 (within the shaded rectangle) being replaced by G_n at these frequencies. In Figure 5.16, the shaded rectangle has been marked as $\tilde{\mathbf{G}}_n$ since only an approximation of $\mathbf{G_n}$ will be achieved up to the bandwidth of the decoupling MIMO disturbance observer which will be the lowest bandwidth in the Q_i-filters in (5.59). Use of $\mathbf{Q(j\omega)} = \mathbf{0}$ at high frequencies results in sensor noise rejection corresponding to the disturbance observer augmented system being replaced by the uncompensated plant $\mathbf{G}$ at these frequencies. This is the desired result which means that the disturbance observer compensated MIMO plant $\mathbf{G}$ can be replaced by the desired or nominal decoupled plant $\mathbf{G_n}$ within the bandwidth of the disturbance observer for the purpose of designing the feedback compensator $\mathbf{C}$ in Figure 5.16.

The disturbance observer will not be able to do decouple loops at frequencies above its individual loop bandwidths. The Q_i-filter bandwidths are, therefore, selected to be above frequencies of significant coupling between the loops. The MIMO decoupled disturbance observer introduced in this section is a straightforward extension of the SISO formulation that enables the designer to reduce the MIMO design problem to several SISO design problems, with loop interactions being treated as disturbances. This is a natural choice for applications where decoupling of the loops in a MIMO plant are desired. This approach works well as the most significant feature of disturbance observer compensation is its ability to reject disturbances within the bandwidth of the Q-filter. Only square MIMO systems can be handled in this approach.

A design based on mapping frequency domain bounds to parameter space can be carried out in a similar manner to the SISO case such that the MIMO system with the MIMO disturbance observer proposed in Section 5.2 is handled loop-at-a-time. In this case, the loop gain is

$$\mathbf{L} = \mathbf{GK} = \mathbf{G[I - Q]^{-1}QG_n}^{-1} \tag{5.66}$$

The equivalent or effective controller K is defined using (5.66) as

$$\mathbf{K} = \mathbf{[I - Q]^{-1}QG_n}^{-1} \tag{5.67}$$

Using (5.58) and (5.59)

$$\mathbf{K} = [\mathbf{I} - \mathbf{diag}(Q_i)]^{-1}\mathbf{diag}(Q_i)\mathbf{diag}\left(\frac{1}{G_{ni}}\right) = \mathbf{diag}\left(\frac{Q_i}{G_{ni}(1 - Q_i)}\right) \text{ for } i = 1, \ldots, m \tag{5.68}$$

The ith element in the diagonal of the equivalent controller $\mathbf{K}$ becomes

$$K_i = K_{Ri} + jK_{Ii} = \frac{Q_i}{G_{ni}(1 - Q_i)} \tag{5.69}$$

which should be solved for two unknown parameters that may be plant uncertainties or controller gains in the parameter space approach.

5.8   Case study: MIMO decoupling disturbance observer control for a four-wheel steering vehicle

A four-wheel steering vehicle (see Figure 5.10) yaw dynamics control is treated in this section on the application of the MIMO decoupling disturbance observer. The underlying single track model and the parameters used were presented earlier in the case study in Section 5.5 and will not be repeated. The two-input–two-output state–space model of this single track four-wheel steering vehicle is

$$\begin{bmatrix} \dot{\beta} \\ \dot{r} \end{bmatrix} = \begin{bmatrix} a_{11} & a_{12} \\ a_{21} & a_{22} \end{bmatrix} \begin{bmatrix} \beta \\ r \end{bmatrix} + \begin{bmatrix} b_{11} & b_{12} \\ b_{21} & b_{22} \end{bmatrix} \begin{bmatrix} \delta_f \\ \delta_r \end{bmatrix}$$

$$\begin{bmatrix} r \\ \beta \end{bmatrix} = \begin{bmatrix} 0 & 1 \\ 1 & 0 \end{bmatrix} \begin{bmatrix} \beta \\ r \end{bmatrix} + \begin{bmatrix} 0 & 0 \\ 0 & 0 \end{bmatrix} \begin{bmatrix} \delta_f \\ \delta_r \end{bmatrix}$$

(5.70)

$$a_{11} = -\frac{c_r + c_f}{\tilde{m}v}, \ a_{12} = -1 + \frac{c_r l_r - c_f l_f}{\tilde{m}v^2}, \ a_{21} = \frac{c_r l_r - c_f l_f}{\tilde{I}_z}, \ a_{22} = -\frac{c_r l_r^2 + c_f l_f^2}{\tilde{I}_z v}$$

$$b_{11} = \frac{c_f}{\tilde{m}v}, \ b_{12} = \frac{c_r}{\tilde{m}v}, \ b_{21} = \frac{c_f l_f}{\tilde{I}_z}, \ b_{22} = -\frac{c_r l_r}{\tilde{I}_z}, \ \tilde{m} = \frac{m}{\mu}, \ \tilde{I}_z = \frac{I_z}{\mu}$$

where all parameters used were explained in Section 5.5. The state–space model (5.70) is a MIMO plant with front and rear wheel steering angles as its two inputs and with yaw rate and side slip angle as its two outputs.

Two uncertain parameters are identified as $\tilde{m}$ and v. They appear polynomially in the coefficients of the transfer functions obtained from the model (5.70). The MIMO decoupling disturbance observer design method in Section 5.7 is used with the four disturbance observer filters G_{n1}, G_{n2}, Q_1, Q_2. The G_{ni} and Q_i-filters are chosen to be of the form given by (5.39) and (5.40) where $K_n = K_n(v)$, the steady-state gain of the single track model at the chosen velocity, is used in (5.39) and loop-at-a-time design was carried out as presented earlier.

The sensitivity function weight is chosen as

$$\mathbf{W}_{\mathbf{S}}^{-1}(s) = \begin{bmatrix} h_{S1} \dfrac{s + \omega_{S1} l_{S1}}{s + \omega_{S1} h_{S1}} & 0 \\ 0 & h_{S2} \dfrac{s + \omega_{S} l_{S2}}{s + \omega_{S2} h_{S2}} \end{bmatrix}$$

(5.71)

with $l_{S1} = 0.5$ (i.e., less than 50% steady-state error) being the low-frequency sensitivity bound, $h_{S1} = 2$ being the high-frequency sensitivity bound and $\omega_{S1} = 25 \, \text{rad/s}$ being the approximate bandwidth of model regulation for the first loop. $l_{S2} = 0.5$, $h_{S2} = 4$ and $\omega_{S2} = 3 \, \text{rad/s}$ were used for the second loop. The complementary sensitivity function weight is chosen as

$$\mathbf{W}_{\mathbf{T}}(s) = \begin{bmatrix} h_{T1} \dfrac{s + \omega_{T1} l_{T1}}{s + \omega_{T1} h_{T1}} & 0 \\ 0 & h_{T2} \dfrac{s + \omega_{T2} l_{T2}}{s + \omega_{T2} h_{T2}} \end{bmatrix}$$

(5.72)

where the low-frequency gain is $l_{T1} = 0.2$ and the high-frequency gain is $h_{T2} = 1.5$ (corresponds to uncertainty of up to 150% at high frequencies). The frequency of transition to significant model uncertainty is $\omega_{T1} = 120 \, \text{rad/s}$ for the first loop. $l_{T2} = 0.2$, $h_{T2} = 2$ with $\omega_{T2} = 50 \, \text{rad/s}$ were used for the second loop. The result with

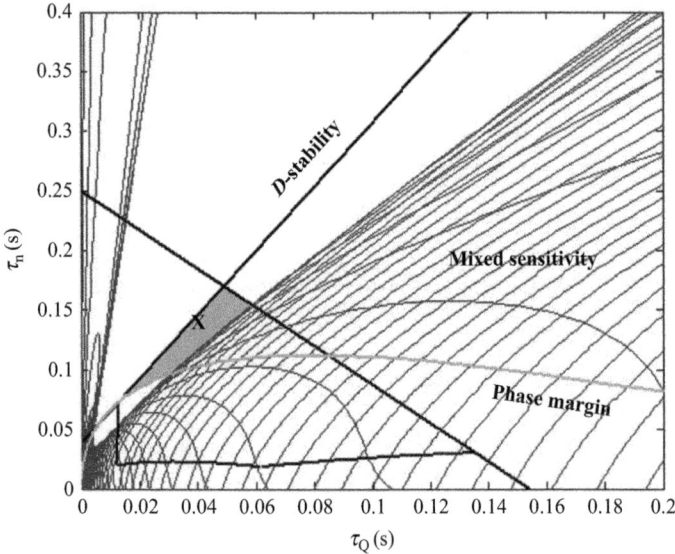

Figure 5.17　Overall solution region for the first loop

this decoupling-type disturbance observer is two separate mixed sensitivity problems given earlier by (5.32) and (5.33). The solution procedure in Section 5.4 is applied to each loop separately and the coupling between the loops is neglected in the analysis as the disturbance observer takes care of this.

The first disturbance observer loop is from the front wheel steering angle to yaw rate and the second disturbance observer loop is from the rear wheel steering angle to vehicle sideslip angle. The cross-coupling effects of front wheel steering angle on vehicle sideslip angle and the rear wheel steering angle on yaw rate are viewed as disturbances to be rejected by the decoupling MIMO disturbance observer, resulting in two independent SISO design procedures for the two loops of the four-wheel car steering system. The overall parameter space solution regions for these two loops for mixed sensitivity, $\mathscr{D}$-stability and a phase margin bound are presented in Figures 5.17 and 5.18. Use of about 30 frequencies is sufficient to get a good characterisation of the overall solution region per objective like $\mathscr{D}$-stability, phase margin constraint, mixed sensitivity, etc.

The chosen disturbance observer filter parameter pairs are marked by crosses in Figures 5.17 and 5.18. The overall solution region for the first disturbance observer loop is obtained by graphical intersection of the solution regions satisfying the mixed sensitivity constraint, $\mathscr{D}$-stability with $\sigma = 4$, $\theta = 60°$ ($\zeta = \cos\theta = 0.5$), $R = \omega_{nmax} = 25\,\text{rad/s}$ and a phase margin greater than 80° as shown in Figure 5.17. The overall solution region for the second disturbance observer loop is obtained by

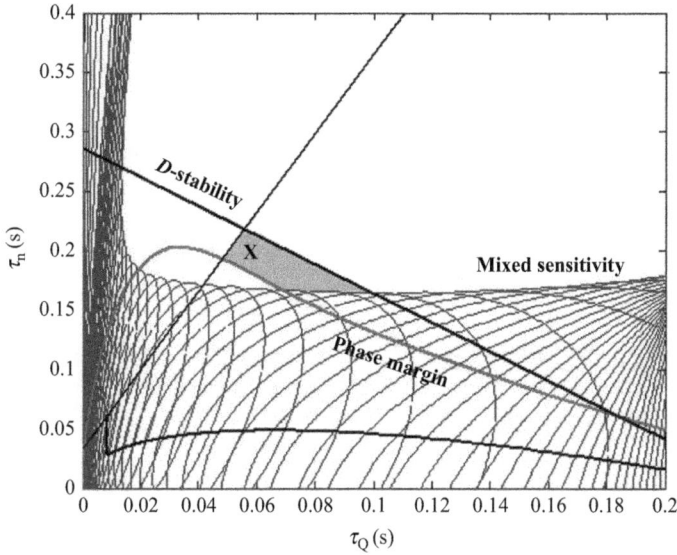

Figure 5.18   Overall solution region for the second loop

graphical intersection of the solution regions satisfying the mixed sensitivity con-
straint, $\mathscr{D}$-stability with $\sigma = 3.5$, $\theta = 60°$ ($\zeta = 0.5$), $R = \omega_{nmax} = 30\,\text{rad/s}$ and a
phase margin greater than $45°$ as shown in Figure 5.18. The resulting control gains
corresponding to the cross signs are $\tau_n = 0.14\,\text{s}$, $\tau_Q = 0.04\,\text{s}$ for the first loop in Figure 5.17 and $\tau_n = 0.20\,\text{s}$, $\tau_Q = 0.06\,\text{s}$ for the second loop in Figure 5.18. These values
are used in the simulations.

Simulations were carried out by inputting step front and rear wheel steering
values. The yaw rate output is shown in Figure 5.19. There is a yaw rate response to
both front and rear wheel steering inputs in the uncontrolled case. Use of the MIMO
decoupling disturbance observer introduced here results in a significantly decoupled
response as the yaw rate responds mainly to the front wheel steering angle and only
negligibly to the rear steering angle in the controlled case. The yaw rate response to
front wheel step steering input behaves like the desired first-order system.

The simulation results from the front and rear steering wheel inputs to the vehicle
side slip angle output are displayed in Figure 5.20. The vehicle side slip angle output
responds mainly to the rear wheel steering input and negligibly to front wheel steering
input in the MIMO decoupling disturbance observer controlled case. Yaw moment
disturbance rejection simulations were also carried out and are shown in Figure 5.21.
Excellent disturbance rejection is achieved with the MIMO decoupling disturbance
observer as neither the yaw rate nor the vehicle side slip angles are affected much by
yaw moment disturbances.

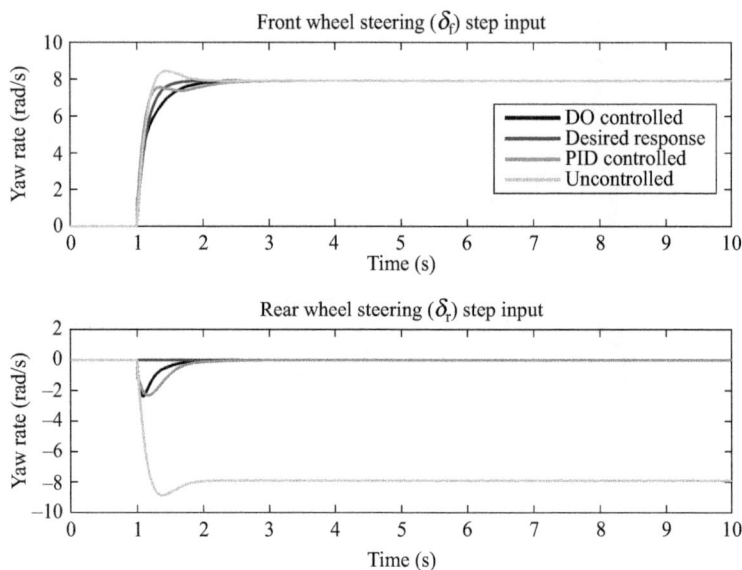

Figure 5.19   Step steering input simulation results for yaw rate output

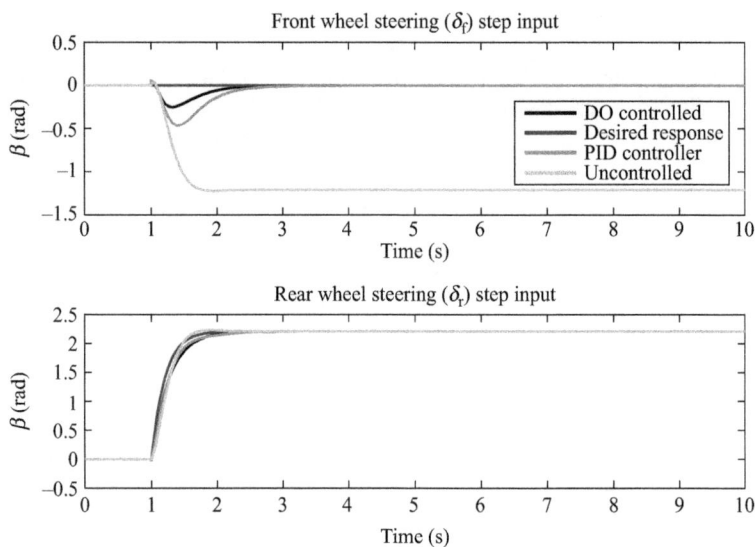

Figure 5.20   Step steering input simulation results for sideslip angle output with sensor noise

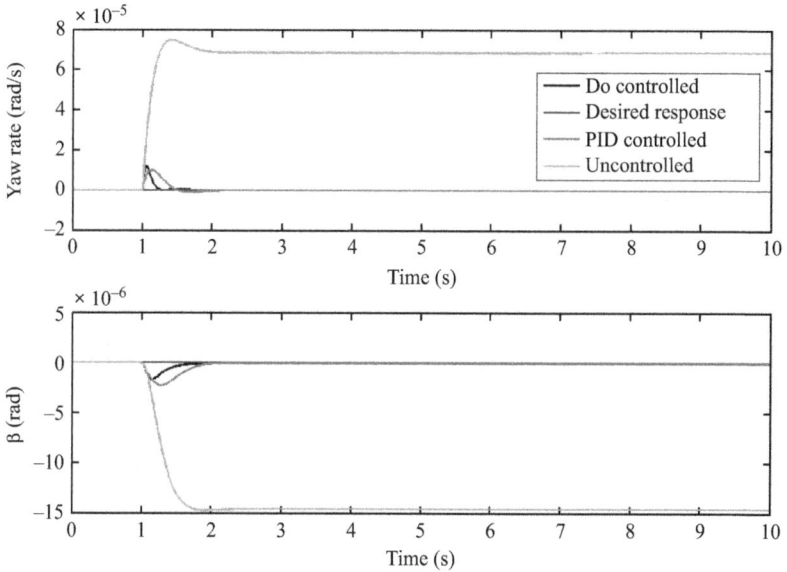

Figure 5.21   Step yaw moment input simulation results for yaw rate and sideslip angle outputs

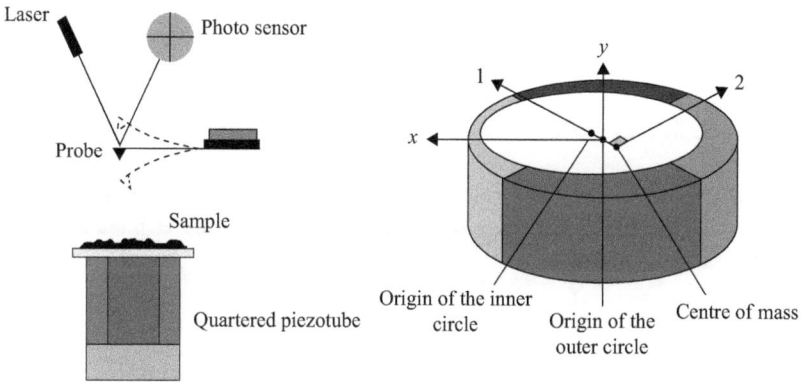

Figure 5.22   Quartered piezotube with eccentricity

5.9   Case study: MIMO disturbance observer for decoupling the two axes of motion in a piezotube actuator used in an atomic force microscope

The MIMO decoupling disturbance observer control method is applied to the lateral plane of scanning motion in a piezoelectric tube based AFM in this case study. Tube shaped piezo-elements having a quartered design like that illustrated in Figure 5.22

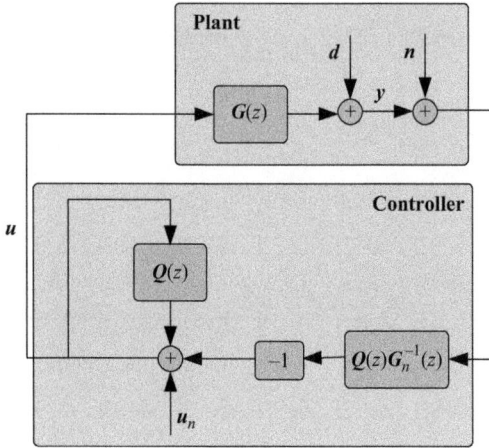

Figure 5.23   MIMO decoupling disturbance observer for piezotube

are commonly used as actuators in cheap AFMs as they can achieve nanometer level resolution. These quartered piezotubes can also be used for nanoscale manipulation tasks. The main problem with these piezotubes is that motion in the x and y lateral plane axes is coupled due to the eccentricity between the inner and outer tubes [7,19,20]. This coupling results in distortion of the scanned image in an AFM application and will result in inaccurate motion in a nanoscale manipulation application. This is an excellent example for the application of the decoupling MIMO disturbance observer.

A discrete-time implementation of the decoupling MIMO disturbance observer of Section 5.8 is used here as illustrated in Figure 5.23. The coupled plant and desired decoupled nominal model are

$$\mathbf{G}(z) = \begin{bmatrix} \dfrac{-0.0057}{z^2 - 1.5017z + 0.9331} & \dfrac{0.0005}{z^2 - 1.5160z + 0.9922} \\[3mm] \dfrac{0.0006}{z^2 - 1.5480z + 0.9596} & \dfrac{0.0044}{z^2 - 1.4995z + 0.9286} \end{bmatrix} \tag{5.73}$$

$$\mathbf{G_n}(z) = \begin{bmatrix} \dfrac{-0.0057}{z^2 - 1.5017z + 0.9331} & 0 \\[3mm] 0 & \dfrac{0.0044}{z^2 - 1.4995z + 0.9286} \end{bmatrix} \tag{5.74}$$

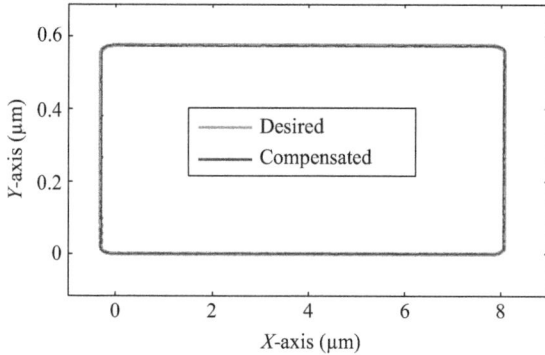

Figure 5.24 Comparison of coupled and decoupling MIMO disturbance observer compensated responses

The **Q** matrix is

$$\mathbf{Q}(z) = \begin{bmatrix} \dfrac{0.247}{z^2 - 1.004z + 0.251} & 0 \\ 0 & \dfrac{0.247}{z^2 - 1.004z + 0.251} \end{bmatrix} \tag{5.75}$$

$\mathbf{u_n}$ in Figure 5.23 is chosen to be a triangular wave on one axis while it is a ramp on the other axis to provide the raster scan motion that is required in an AFM scanning application. The coupled rectangular desired motion $\mathbf{u_n}$ in Figure 5.24 is used to demonstrate the effectiveness of the decoupling MIMO disturbance observer of this case study. While the coupled model in (5.74) results in a skewed rectangle in the simulations, the decoupled MIMO disturbance observer forces it to behave like the decoupled model in (5.74), resulting in the perfect rectangular motion profile in Figure 5.24. The reader is referred to reference [7] for more detailed information and for experimental results.

5.10 Communication disturbance observer

Control of plants with time delay is a difficult procedure due to the destabilising effect of the time-delay [21,22]. The conventional approach to time delay compensation is to use the Smith predictor [23] which has been extended for various cases in references [24,25]. The Smith predictor and its extensions are simple to understand and are easily implemented. However, the use of a Smith predictor requires that the amount of time delay is known exactly with any inaccuracy in its knowledge resulting in degraded performance.

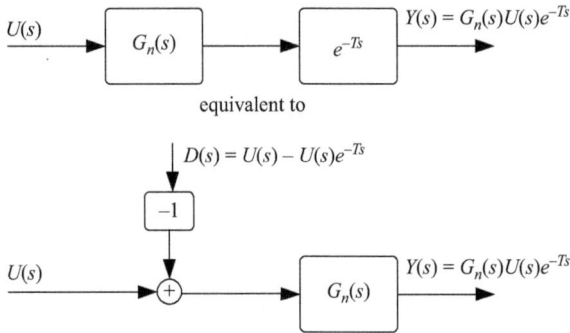

Figure 5.25   Network disturbance concept

A compensation method called the communication disturbance observer was introduced to compensate for the time delay inherent in bilateral teleoperation systems by Natori *et al.* [26]. It can also be applied to plants with variable time delays. The structure of the communication disturbance observer is similar to the structure of the disturbance observer except the disturbance definition and time delay compensation parts. The time delayed system is re-written using the network disturbance concept. The effect of the time delay can be expressed as shown in Figure 5.25 where the time delay is seen as a disturbance that is acting on the system.

The network disturbance is defined as

$$d\,(t) = u\,(t) - u\,(t - T) \tag{5.76}$$

or by after applying the Laplace transform as

$$D\,(s) = U\,(s) - U\,(s)\,e^{-Ts} \tag{5.77}$$

where u is the system input and T is the time delay. $D(s)$ in (5.77) is defined as the network disturbance.

The communication disturbance observer and its equivalent sequential block diagram transformations are shown in Figures 5.26 and 5.27, respectively. The communication disturbance observer structure consists of two parts: (i) network disturbance estimation and (ii) time delay compensation. The communication disturbance observer estimates the network disturbance which is used to compensate for the time delay effect in the feedback signal.

As our knowledge of the nominal plant G_n may have uncertainty in it, the nominal plant in the top path in Figure 5.26 is better represented by $G = G_n(1 + \Delta_m)$ where Δ_m represents the multiplicative uncertainty in our knowledge of the plant. With this change, the closed-loop system transfer function is given by

$$\frac{y}{r} = \frac{CGe^{-Ts}}{1 + CG_nQ + CGe^{-Ts}\,(1 - Q)} \tag{5.78}$$

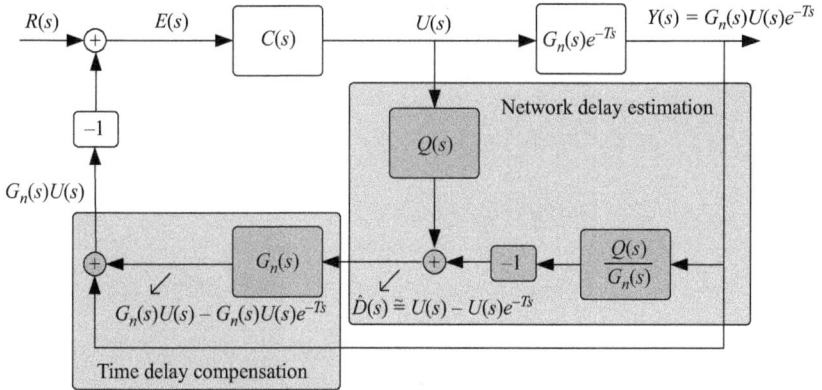

Figure 5.26 Communication disturbance observer

The choice of Q as a unity gain low-pass filter will result in the elimination of the time delay term in the denominator of the transfer function in (5.78) as

$$\frac{y}{r} \cong \frac{CGe^{-Ts}}{1 + CG_nQ} \text{ for } Q \to 1 \tag{5.79}$$

and the time delay is compensated for by the communication disturbance observer. Obviously, the output is still delayed by the time delay like that in the Smith predictor but unlike the Smith predictor, we do not need to know what the time delay T is. The design of the communication disturbance observer is very similar to that of the disturbance observer itself. We still need to choose the G_n and Q-filters. The bandwidth of the Q-filter determines the magnitude of time delay that can be accommodated. Large time delay values will require a higher bandwidth of the Q-filter which will eventually be limited by the actuator bandwidth of the controlled system. Once the communication disturbance observer is designed and implemented, a feedback controller can be designed for the nominal plant G_n without any time delay.

For a feedback control system with controller K, the loop transfer function of the system with multiplicative uncertainty is

$$L = KG = KG_n(1 + \Delta_m) = L_n + L_n\Delta_m \tag{5.80}$$

where $L_n = KG_n$ is the nominal loop transfer function. If we assume that the nominal closed-loop system is stable, robust stability of the uncertain system can be guaranteed if L does not encircle the point $(-1, 0)$ according to Nyquist stability criterion or

$$|\Delta_m(j\omega)L_n(j\omega)| < |1 + L_n(j\omega)|, \quad \forall\omega \tag{5.81}$$

or equivalently

$$\left|\frac{\Delta_m(j\omega)L_n(j\omega)}{1 + L_n(j\omega)}\right| < 1, \quad \forall\omega \Leftrightarrow \|\Delta_m T_n\|_\infty < 1 \tag{5.82}$$

where T_n is the nominal complementary sensitivity function [27, 28].

Using the equivalent system in Figure 5.27 part (c) and incorporating the feedback loop term into the controller, the controller K for the communication disturbance observer-based compensator becomes

$$K = \frac{C(1-Q)}{1+CG_nQ} \tag{5.83}$$

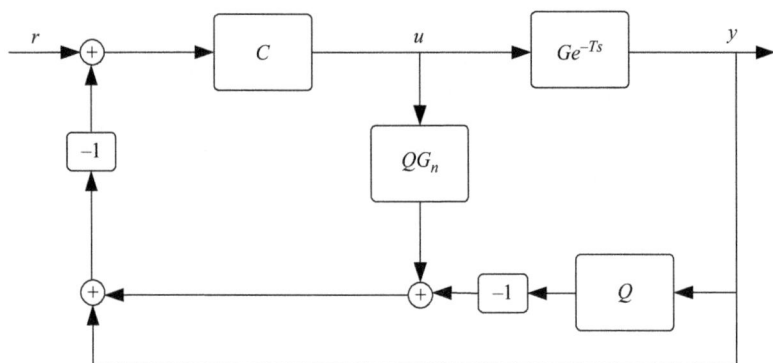

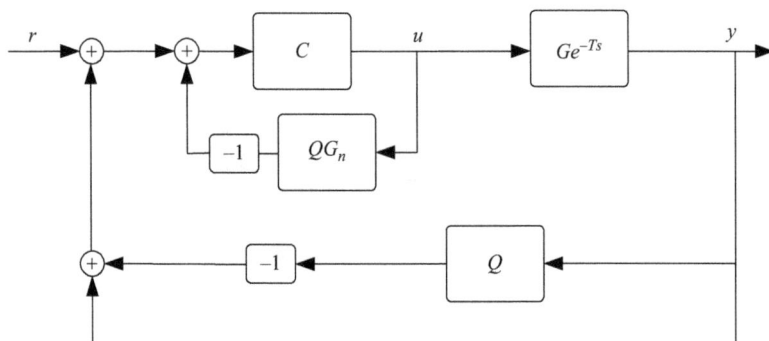

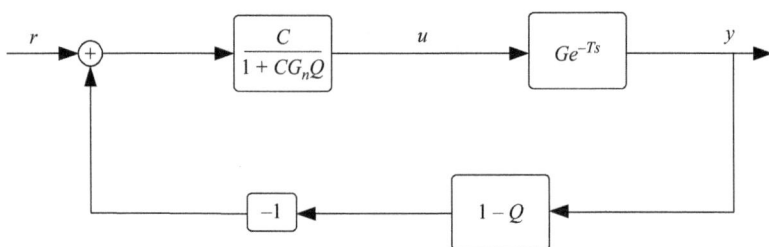

(a)

(b)

(c)

Figure 5.27   Equivalent block diagram transformations of communication disturbance observer

Then, the nominal loop transfer function in (5.80) becomes

$$L_n = KG_n e^{-Ts} = \frac{C(1-Q)\,G_n e^{-Ts}}{1 + CG_n Q} \tag{5.84}$$

Using (5.84), the nominal complementary sensitivity transfer function is given by

$$T_n = \frac{L_n}{1+L_n} = \frac{CG_n e^{-Ts}(1-Q)}{1 + CG_n Q + CG_n e^{-Ts}(1-Q)} \tag{5.85}$$

As a result, the robust stability condition in (5.82) becomes

$$\left| \frac{CG_n e^{-Ts}(1-Q)}{1 + CG_n Q + CG_n e^{-Ts}(1-Q)} \right| < \left| \frac{1}{\Delta_m} \right|, \quad \forall \omega \tag{5.86}$$

for the communication disturbance observer-based controlled system.

The communication disturbance observer can also be used when the time delay is varying. In the case of time varying delay, the robust stability condition in (5.86) becomes

$$\left| \frac{CG_n e^{-T_{max}s}(1-Q)}{1 + CG_n Q + CG_n e^{-T_{max}s}(1-Q)} \right| < \left| \frac{1}{e^{-T_{max}s} - 1} \right|, \quad \forall \omega \tag{5.87}$$

The unity gain low-pass Q-filter can be designed considering the restriction of the time delay upper bound T_s with some conservatism [[27],28].

5.11   Case study: communication disturbance observer application to vehicle yaw stability control over CAN bus

Time delay occurs and causes stability problems in many automotive control systems such as idle speed control, air-to-fuel ratio control, anti-jerk control, cooperative adaptive cruise control and other controller area network (CAN) based distributed control systems in a vehicle. Figure 5.28 shows the basic elements of a typical networked control system in a road vehicle. There are two network-induced delays: controller to

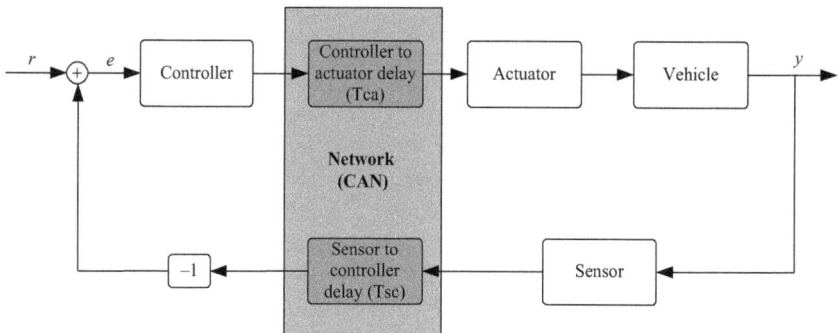

Figure 5.28   The basic elements of a networked control system in a road vehicle

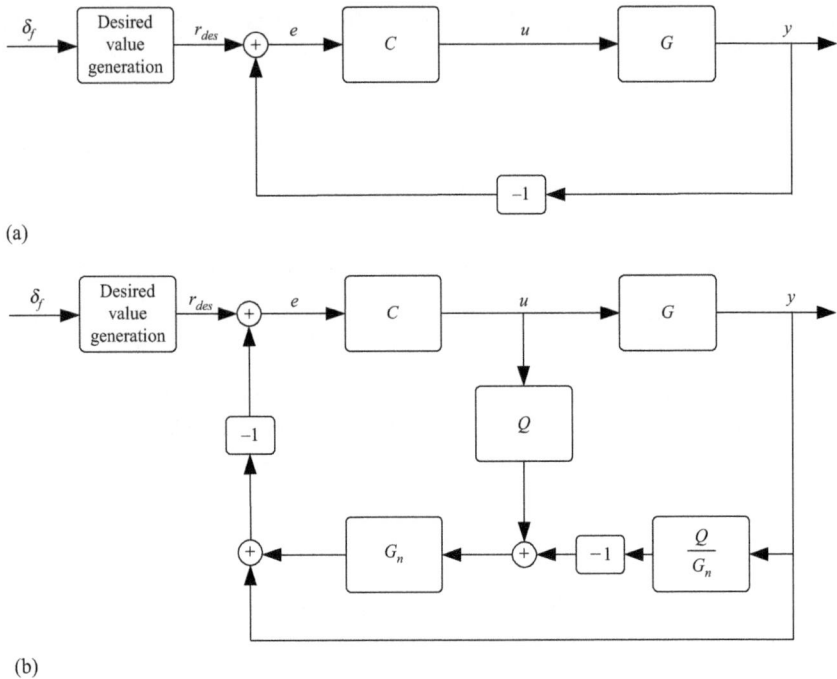

(a)

(b)

Figure 5.29   The yaw stability control system block diagrams (a) with only feedback control and (b) with communication disturbance observer and feedback control

actuator delay (T_{ca}) and sensor to controller delay (T_{sc}). In this section, the communication disturbance observer approach is used to compensate for the network-induced delays in the yaw stability control system of the vehicle [28].

Figure 5.29 shows the yaw stability control systems with only feedback control and with the communication disturbance observer and feedback control. In yaw stability control, the aim is to regulate the yaw rate to its desired value which is calculated using the front wheel steering angle input and single track vehicle model in desired value generation block in Figure 5.29. The yaw rate error e obtained is used by the feedback controller C to generate the controller signal u. Here, the feedback controller C is a PI controller. The single track model, its parameters and numerical values used were introduced earlier in Section 5.5. The time varying network-induced delay in the CAN bus consists of the sum of the controller to actuator delay T_{ca} and the sensor to the controller delay T_{sc} with the total time varying delay varying between 6 ms and 20 ms [29]. Figure 5.30 shows the bounded time-varying delay $T(t)$ used in the simulations.

The unity gain low-pass Q-filter can be designed considering the robust stability condition for the bounded time varying delay given in (5.87). This condition is shown

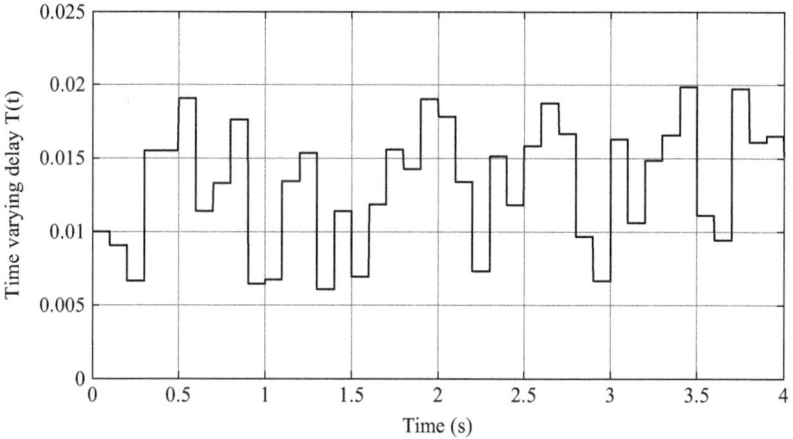

Figure 5.30   Time varying delay T(t) in CAN communication

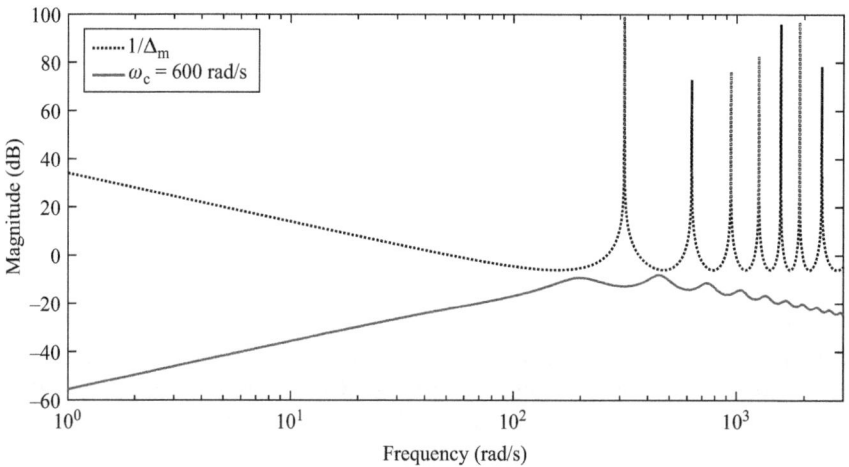

Figure 5.31   Stability of robustness for the time varying delayed yaw stability control system with communication disturbance observer

in Figure 5.31 for $Q = \omega_c/(s + \omega_c)$ with a cut-off frequency of 600 rad/s. It is seen that the uncertainty and robust stability lines do not intersect each other for $\omega_c = 600\,\text{rad/s}$. The feedback controlled system is stable for the selected cut-off frequency.

A simulation with front wheel steering input magnitude of 0.14 rad is conducted to at the constant vehicle velocity of 30 m/s to test the effectiveness of the communication disturbance observer. The tyre–road friction coefficient μ is selected as unity for dry road conditions. The desired yaw rate and the yaw rate responses of the PI feedback controlled vehicle and the PI feedback plus communication disturbance

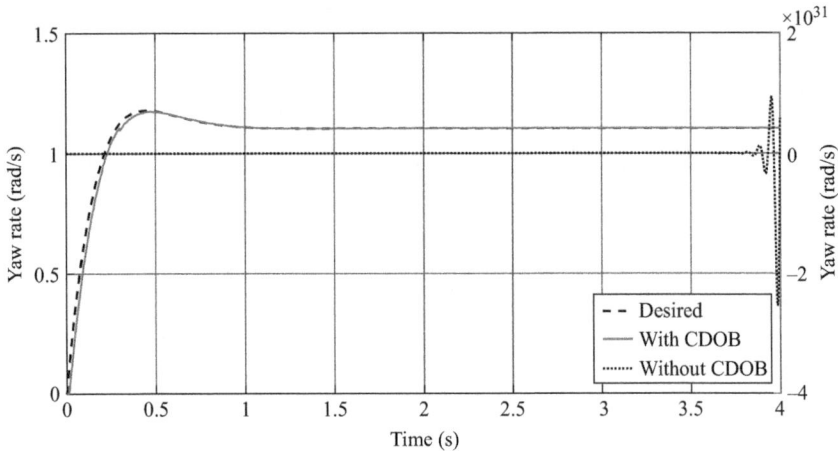

Figure 5.32   Simulation results for the yaw stability control system with and without communication disturbance observer

observer controlled vehicle are shown in Figure 5.32. It is seen that the time varying delay causes instability in the case of PI feedback control alone. However, in the case of the communication disturbance observer and feedback control, the vehicle yaw rate response is stable and follows the desired yaw rate successfully.

5.12   Chapter summary and concluding remarks

The disturbance observer was introduced in this chapter as a two-degrees-of-freedom control architecture that regulates a plant with uncertainty around a desired model while providing excellent disturbance rejection. Discrete-time, decoupling MIMO and time delay compensating extensions of the disturbance observer were presented. The basic design consisted of choosing a desired response transfer function and a unity d.c. gain low-pass Q-filter. Parameter space based design was also introduced and used. Several different case studies from vehicle yaw dynamics control, electromechanical system motion control and piezotube control were presented to demonstrate the effectiveness of the disturbance observer method.

References

[1]   K. Ohnishi, "A new servo method in mechatronics," *Transactions of the Japan Society of Electronics Engineering*, vol. 107, pp. 83–86, 1987.

[2]   T. Umeno and Y. Hori, "Robust speed control of DC servomotors using modern two degrees-of-freedom controller design," *IEEE Transactions on Industrial Electronics*, vol. 38, no. 5, pp. 363–368, Oct. 1991.

[3]   L. Güvenç and K. Srinivasan, "Friction compensation and evaluation for a force control application," *Mechanical Systems and Signal Processing*, vol. 8, no. 6, pp. 623–638, 1994.

[4]   C. J. Kempf and S. Kobayashi, "Disturbance observer and feedforward design for a high-speed direct-drive positioning table," *IEEE Transactions on control systems Technology*, vol. 7, no. 5, pp. 513–526, Sept. 1999.

[5]   B. Aksun-Güvenç, L. Güvenç, and S. Karaman, "Robust yaw stability controller design and hardware-in-the-loop testing for a road vehicle," *IEEE Transactions on Vehicular Technology*, vol. 58, no. 2, pp. 555–571, Feb. 2009.

[6]   B. Aksun-Güvenç, L. Güvenç, and S. Karaman, "Robust MIMO disturbance observer analysis and design with application to active car steering," *International Journal of Robust and Nonlinear Control*, vol. 20, no. 8, pp. 873–891, 2010.

[7]   B. Aksun-Güvenç, S. Necipoğlu, B. Demirel, and L. Güvenç, *Robust Control of Atomic Force Microscopy*, ser. Mechatronics. London: ISTE/Wiley, March 2011, ch. 3.

[8]   X. Fan and M. Tomizuka, "Robust disturbance observer design for a power-assist electric bicycle," in *Proceeding of the American Control Conference*, 2010.

[9]   T. Satoh, K. Kaneko, and N. Saito, "Improving tracking performance of predictive functional control using disturbance observer and its application to table drive systems." *International Journal of Computers, Communications & Control*, vol. 7, no. 3, 2012.

[10] Y. Luo, T. Zhang, B.-J. Lee, C. Kang, and Y.-Q. Chen, "Disturbance observer design with Bode's ideal cut-off filter in hard-disc-drive servo system," *Mechatronics*, vol. 23, no. 7, pp. 856–862, 2013.

[11] T. Bünte, D. Odenthal, B. Aksun-Güvenç, and L. Güvenç, "Robust vehicle steering control design based on the disturbance observer," *Annual Reviews in Control*, vol. 26, no. 1, pp. 139–149, 2002.

[12] B. Aksun-Güvenç, T. Bünte, D. Odenthal, and L. Güvenç, "Robust two degree-of-freedom vehicle steering controller design," *IEEE Transactions on Control Systems Technology*, vol. 12, no. 4, pp. 627–636, July 2004.

[13] B. Aksun-Güvenç, "Applied robust motion control," Ph.D. dissertation, Istanbul Technical University, Istanbul, Turkey, 2001.

[14] L. Güvenç and J. Ackermann, "Links between the parameter space and frequency domain methods of robust control," *International Journal of Robust and Nonlinear Control*, vol. 11, no. 15, pp. 1435–1453, Dec. 2001.

[15] J. Ackermann, P. Blue, T. Bünte, L. Güvenç, D. Kaesbauer, M. Kordt, M. Mühler, and D. Odhental, *Robust Control: The Parameter Space Approach*. Springer Verlag: London, UK, 2002.

[16] C.-H. Menq and K.-C. Hsia, "Discrete model regulation for systems with uncertain dynamics," in *Proceedings of American Control Conference*, 1992.

[17] K. C. Hsia, "Sensitivity reduction and disturbance rejection via discrete model regulation," Ph.D. dissertation, Ohio State University, Columbus, OH, USA, 1993.

[18]   S. Skogestad and I. Postlethwaite, *Multivariable feedback control: analysis and design.* Wiley New York, 2007, vol. 2.

[19]   S. O. R. Moheimani, "Invited review article: Accurate and fast nanopositioning with piezoelectric tube scanners: Emerging trends and future challenges," *Review of Scientific Instruments*, vol. 79, no. 7, p. 071101, 2008.

[20]   S. Devasia, E. Eleftheriou, and S. O. R. Moheimani, "A survey of control issues in nanopositioning," *IEEE Transactions on Control Systems Technology*, vol. 15, no. 5, pp. 802–823, Sept. 2007.

[21]   J.-P. Richard, "Time-delay systems: An overview of some recent advances and open problems," *Automatica*, vol. 39, no. 10, pp. 1667–1694, Oct. 2003.

[22]   K. Gu and S.-I. Niculescu, "Survey on recent results in the stability and control of time-delay systems," *ASME Journal of Dynamic Systems, Measurement and Control*, vol. 125, no. 2, pp. 158–165, June 2003.

[23]   O. J. Smith, "A controller to overcome dead time," *ISA Journal*, vol. 6, no. 2, pp. 28–33, 1959.

[24]   A. S. Rao and M. Chidambaram, "Enhanced Smith predictor for unstable processes with time delay," *Industrial & Engineering Chemistry Research*, vol. 44, no. 22, pp. 8291–8299, Sept. 2005.

[25]   J. E. Normey-Rico and E. F. Camacho, "Unified approach for robust dead-time compensator design," *Journal of Process Control*, vol. 19, no. 1, pp. 38–47, Jan. 2009.

[26]   K. Natori, T. Tsuji, K. Ohnishi, A. Hace, and K. Jezernik, "Robust bilateral control with internet communication," in *Proceedings of the 30th Annual Conference of IEEE Industrial Electronics Society*, 2004.

[27]   M. T. Emirler, B. Aksun-Güvenç, and L. Güvenç, "Communication disturbance observer approach to control of integral plant with time delay," in *Proceeding of pth IEEE Asian Control Conference*, 2013.

[28]   M. T. Emirler, "Advanced control systems for ground vehicles," Ph.D. dissertation, Istanbul Technical University, Istanbul, Turkey, 2015.

[29]   C. Latrach, M. Kchaou, A. Rabhi, and A. El-Hajjaji, "H_∞ networked fuzzy control for vehicle lateral dynamic with limited communication," *IFAC Proceedings Volumes*, vol. 47, no. 3, pp. 6313–6318, 2014.

Chapter 6

Repetitive control

Repetitive control uses a time delay element in a feedback loop to reduce tracking error for systems with periodic reference or disturbance inputs of known period. The SISO continuous-time and discrete-time repetitive control architectures and design methods based on minimising mixed sensitivity at the fundamental periodic frequency and its harmonics are introduced in this chapter. The chapter also has COMES toolbox design examples. The first example is on an AFM application while the second example is on a Quanser QUBE™ servo system application.

6.1 Introduction to repetitive control

There is a group of applications where the control problem addressed has periodic reference inputs or periodic disturbances. Controllers that take into account the periodic nature of the exogenous input work much better for such applications. Repetitive control is such a control scheme that has been developed specifically for handling systems with repetitive references/disturbances and will be treated in this chapter. Its architecture is shown in the block diagram of Figure 6.1 as a plug-in repetitive control system. It is plug-in because it is added to the main feedback control system as an extra loop. If it is turned off, the main feedback loop will continue to work if the plant already includes compensation. Repetitive control reduces error in systems with periodic reference inputs or disturbances with known period by introducing a highly frequency selective gain through a positive feedback loop containing a time delay element. This positive feedback loop is a generator of periodic signals. The delay time is equal to the known period of the periodic reference/disturbance. Significant improvements in the tracking accuracy or disturbance rejection characteristics of systems subject to periodic exogenous inputs can thus be achieved.

The repetitive control system is a special type of servo-system whose basic structure is based on the Internal Model Principle of Francis and Wonham [1]. The idea of repetitive control was first created by Inoue *et al.* [2] to replace conventional motion control techniques in the control of a proton synchrotron magnet power supply. It has been widely utilised in many application areas including control of hard disc drives [3], control of optical disc drives [4], control of non-circular tuning [5], trajectory control of industrial robot arms [6–9], motor speed control [10], high precision rotational control [11–13], control of material testing machine [14], control of cold rolling process [15], suppression of torque vibration in motors [16,17], control of active air bearing [18], acoustic impedance matching [19], reduction of waveform distortion in PWM inverter or UPS [20–22], accurate position control of piezoelectric actuators [23,24] and control of the piezo-stages in AFMs [25–28].

The earlier papers in the literature have generally focused on the stability analysis in both continuous-time [29–31] and discrete-time systems [32,14]. Srinivasan and Shaw [31] have created an analysis tool called the *regeneration spectrum* to study the absolute and relative stability of repetitive control systems by estimating the locus of the dominant closed-loop poles. Their results have also provided an improved insight into design trade-offs between performance and stability. Tsao and Tomizuka [5] have analysed the robust stability of repetitive control systems applied to plants with unstructured modelling error. In order to achieve a specified level of nominal performance, Srinivasan *et al.* [33] have utilised the Nevanlinna-Pick interpolation method to modify repetitive controller design by means of optimising a measure of stability robustness. Peery and Özbay have modified $\mathscr{H}_\infty$ optimal design approach presented in [34] and then applied the extension of this methodology based on Youla parameterisation to repetitive control systems in [35]. Moon *et al.* [4] have developed a robust design methodology for parametric uncertainty in interval plants under repetitive control. Similarly, Roh and Chung [36] have created a new synthesis method based on Kharitonov's theorem for repetitive control systems with uncertain parameters. Weiss *et al.* [37–39] have made a stability and robustness analysis for MIMO repetitive control systems based on $\mathscr{H}_\infty$ control theory. μ analysis has been used for assessing stability and performance robustness of SISO continuous-time repetitive control systems by Güvenç [40]. μ synthesis has been applied to sampled data repetitive controller design by Li and Tsao [41]. The authors of [42–44] have proposed a continuous-time parameter space procedure to design repetitive controllers satisfying a robust performance criterion. Apart from the aforementioned works, Doh and Chung [45] developed a repetitive control design methodology to guarantee the robust stability of linear systems with time-varying uncertainties.

The repetitive controller design approach, presented in this book, relies on mapping nominal performance, robust performance, and robust stability frequency domain constraints to controller parameter space. In fact, the repetitive controller design approach introduced in this book is significantly different from those of the aforementioned references including the application of $\mathscr{H}_\infty$ methods. The significant advantages of the approach here in comparison with $\mathscr{H}_\infty$ methods are: (i) the ease of visualisation due to the graphical representation of the solution in the parameter space approach and the capability and ease of doing multi-objective optimisation by simply intersecting solution regions for different objectives, (ii) the determination of a solution region rather than one specific solution for the control system satisfying a frequency domain constraint (this makes it easier to design non-fragile controllers as changes in controller parameters will not violate the chosen objectives so long as the parameters are within the solution region), (iii) the determination of controller parameters that guarantee robust performance, (iv) being able to treat plants with time delay and poles on the imaginary axis, (v) not having to use rational, continuous weights in the robust performance specifications and (vi) obtaining fixed structure low-order repetitive controller filters that are easily implementable. There are also some shortcomings of the proposed design method in comparison to the methods that exist in the literature including $\mathscr{H}_\infty$ methods such as: (i) the method can simultaneously accommodate the design of only two controller parameters due to its graphical display of the solution region; (ii) the method does not result in a single analytical solution and the methods used do not look mathematically elegant as a constructive frequency-by-frequency design approach is used.

It is challenging to apply standard robust control methods like $\mathscr{H}_\infty$ control to repetitive controller design for robust performance as the repetitive control system is infinite dimensional due to the presence of the inherent time delay in the controller. Robust control methods such as $\mathscr{H}_\infty$ optimal control have been extended to infinite dimensional systems and applied to repetitive control; see, e.g., [33,35]. However, very high order weighting functions need to be used in the robust controller synthesis. Consequently, the resulting repetitive controller filters also have high order. Model order reduction techniques are used to reduce the order of the repetitive controller filters in an actual implementation. Some of the most powerful characteristics of the proposed method are that the weights utilised in the design do not need to be continuous functions of frequency and that plants can have a time delay and/or poles on the imaginary axis because the computations are naturally carried out only at the frequencies of interest. Secondly, the choice of the frequency grid used is not a problematic issue for the repetitive control design procedure presented here as the main design frequencies are exactly known and are the fundamental frequency of the periodic exogenous signal (reference or disturbance) with the period τ_d and its harmonics. The largest harmonic frequency considered is chosen to be close to the bandwidth of the repetitive control system which is limited by the bandwidth of the actuator used in the implementation. The method presented here is for SISO systems; however, it can also be used to design controllers for MIMO systems where one loop at a time design is possible.

The organisation of the rest of chapter is as follows. In Section 6.2, some analysis techniques on the stability of repetitive control system are presented. Section 6.3 gives basic information on the design principles of robust repetitive control systems. In Section 6.4, the technique of mapping mixed sensitivity frequency domain specifications into repetitive controller parameter space is explained in detail. A numerical example of high-speed AFM scanner position control application is utilised in order to demonstrate the effectiveness of the proposed method presented in Section 6.5. A case study using the Quanser QUBE™ servo system is presented in Section 6.6. In Section 6.7, the repetitive control system part of the COMES toolbox is demonstrated. Extension of the method to discrete-time repetitive control is presented in Section 6.8. The chapter ends with conclusions in Section 6.10.

6.2 Stability analysis of repetitive control system

Stability analysis is a pre-requisite for successful control system design. Well-known analysis techniques, such as the Nyquist stability theorem or the root-locus plot, can

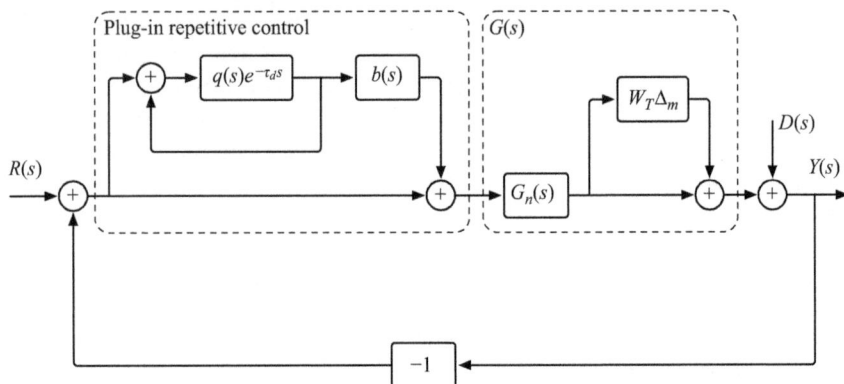

Figure 6.1   Plug-in repetitive control structure

be used to check the stability of a conventional controller for a continuous-time, SISO, LTI plant. In the parameter space approach to robust controller design, the regions of the parameter space where Hurwitz stability and the $\mathscr{D}$-stability conditions are met are determined to guarantee the absolute stability and relative stability, respectively. These methods were presented for conventional control systems in Chapter 2.

In contrast to the stability analysis of conventional (finite-dimensional) control systems, stability analysis of repetitive control systems is difficult to handle due to the presence of the time delay in the controller. These systems are called infinite-dimensional as they have infinitely many poles. The stability analysis and controller design for infinite-dimensional systems are complicated as conventional techniques are not directly applicable; see, e.g., [46,47]. In early applications of repetitive control, the two filters $q(s)$ and $b(s)$, shown in Figure 6.1, were not used in the formulation and the repetitive controller was just the positive feedback loop around the time delay element introduced into the system block diagram. This, however, created stability and robustness problems and the filter $q(s)$, seen in Figure 6.1, was added to shut down repetitive control at frequencies above its bandwidth. The other filter $b(s)$, given in Figure 6.1, is used as an extra degree of freedom in designing the repetitive controller for improved accuracy. Some references use $b(s) = 1$ and do not rely on this additional degree of freedom.

A brief synopsis of the early history on the stability analysis of repetitive control systems from the literature is presented next. In the earlier papers, Inoue *et al.* [2] used the BIBO stability criterion. The analysis given in [48] exposed that stability might be guaranteed only if the plant was proper but not strictly proper. From this analysis, it was understood that there is a significant limitation due to controller parameterisation [49] as well as an unrealistic demand for perfect tracking of arbitrary periodic signals that include high-frequency components. To tackle the problem mentioned above, Inoue *et al.* [2] modified the repetitive controllers by adding the low-pass filter $q(s)$ in front of the controller time delay. As a consequence, the stability robustness of repetitive control systems was improved; however, the tracking performance was dramatically reduced at high frequencies. Indeed, this is an expected situation since the addition of the low-pass filter $q(s)$ in front of the time delay causes the sensitivity of control system to be greater than zero at the fundamental frequency and its harmonics. Therefore, the tracking performance decreases. Hara *et al.* [29] extended the repetitive controller to MIMO systems and presented the corresponding exponential stability. For SISO continuous-time repetitive control systems, Hara *et al.* indicated that any low-pass filter $q(s)$ satisfying $\|q(s)\|_\infty < 1$ can be utilised in minimum phase systems [48]. However, there are bandwidth limitations on $q(s)$ for non-minimum phase systems [50].

There are many different design methods in the literature for repetitive control systems. The design problem is essentially based upon choosing and optimising the low-pass filter $q(s)$ and dynamic compensator $b(s)$ for the plug-in repetitive control system seen in Figure 6.1. There are other architectures for a repetitive control implementation, but the one in Figure 6.1 will be used here. The results presented in this chapter can easily be extended to other repetitive control architectures. The selection of repetitive controller parameters involves a trade-off between steady-state accuracy and stability robustness. $G(s)$ is the plant which may already include a compensator. The plant is the nominal plant $G_n(s)$ with multiplicative uncertainty $W_T \Delta_m$. If the repetitive control loop is turned off ($q(s) = 0$ or $b(s) = 0$), the closed loop transfer function from the command input $R(s)$ to the output $Y(s)$ will be $G_{cl}(s) = G(s)/(1 + G(s))$. The low-pass filter $q(s)$ and dynamic compensator $b(s)$ must be chosen such that $\|q(s)(1 + b(s)G_{cl}(s))\|_\infty < 1$ to satisfy a sufficient condition for stability.

There are some pole assignment methodologies based on the parameter space approach in the literature that can be used to design a controller for continuous-time, LTI, time-delayed systems; see, e.g., [51]. However, these methods are not well-suited for repetitive control systems. Frequency domain methods offer the best approach for repetitive control system design because the use of the pole assignment approach for repetitive control systems is tough due to the presence of the controller time-delay. For these reasons, shaping a function of frequency called the regeneration spectrum which is related to relative stability and accuracy is used here as an effective way of improving the relative stability and transition response of the system. In the following, the regeneration spectrum analysis will be given in detail along with its use in repetitive control system design.

6.2.1 Regeneration spectrum analysis

The characteristic equation of the continuous-time, single-input–single-output system with single time-delay $\tau_d \in \{\tau \in \mathbb{R}_+ : \tau < \infty\}$ is given by

$$P(s) + Q(s)e^{-\tau_d s} = 0, \tag{6.1}$$

where $P(s)$ and $Q(s)$ are polynomials in s. To analyse the stability of such a system, the regeneration spectrum, i.e., $\mathscr{R} : \mathbb{C} \to \mathbb{R}_+$, is defined as

$$\mathscr{R}(\omega) = \left| \frac{Q(j\omega)}{P(j\omega)} \right|. \tag{6.2}$$

The term *regeneration spectrum* is borrowed from the machine tool chatter analysis literature [52]. Equation (6.1) has infinitely many roots, which can be only computed numerically, resulting in a considerable computational effort. The regeneration spectrum proposes an alternative way to estimate the dominant roots of the characteristic equation (6.1) and to obtain a good estimate of the dominant features of the system's dynamic response. Srinivasan and Nachtigal [52] demonstrated that the real parts of the roots of (6.1) can be approximately computed by

$$\alpha_{i,-i} = \text{Re}\{s_{i,-i}\} = \frac{\ln(\mathscr{R}(\omega_i))}{\tau_d}, \tag{6.3}$$

where $s_{i,-i}$ are complex conjugate roots with their imaginary parts equal to ω_i. The continuum of complex numbers $\ln(\mathscr{R}(\omega))/\tau_d \pm j\omega$ for $\omega \in [0, \infty)$ contains the dominant roots of the characteristic equation. The spacing of the roots on the imaginary axis is approximately equal to $2\pi/\tau_d$. Consequently, these roots are closely spaced for large values of the time delay τ_d. The real part of the dominant (stable) characteristic root can be approximately calculated by

$$\alpha_{\max} = \max_i \{\text{Re}\{s_{i,-i}\}\} \cong \max_{0 \leq \omega < \infty} \frac{\ln(\mathscr{R}(\omega))}{\tau_d}. \tag{6.4}$$

Equation (6.4) approximates the real part of the roots of (6.1) expressed as

$$\text{Re}\{s_{i,-i}\} = \frac{\ln\left|\frac{Q(\alpha_i + j\omega_i)}{P(\alpha_i + j\omega_i)}\right|}{\tau_d}. \tag{6.5}$$

Notice that (6.4) can be obtained by setting $\alpha_i = 0$ in (6.5). For the validity of the approximation mentioned above, the following conditions must be satisfied:

(i)   The roots of the characteristic equation of the non-time-delayed system (i.e., $\tau_d \to \infty$),

$$P(s) = 0, \tag{6.6}$$

must be in the left-half of the complex plane.

(ii)   The time delay τ_d needs to satisfy

$$\tau_d \geq \frac{5}{|\alpha_{\max}|} \tag{6.7}$$

where $\alpha_{\max}$ is the real part of the dominant stable root of (6.6).

6.2.2   *Regeneration spectrum analysis applied to repetitive control*

As mentioned earlier, the basic repetitive control system must be improved by using a low-pass filter $q(s)$ in front of the time delay located in the periodic signal generator so that stability can be guaranteed at high frequencies above the bandwidth of $q(s)$. However, the tracking performance at high frequencies is sacrificed by adding a low-pass filter $q(s)$. The continuous-time SISO repetitive control system model, given in Figure 6.1, is very similar to the control system configuration proposed by Hara *et al.* [48]. The difference between the two control system configurations is the use of the dynamic compensator $b(s)$ whose effect on system stability will be presented below.

The transfer function of the repetitive controller $C(s)$ within the dashed box in Figure 6.1 is

$$C(s) = 1 + \frac{b(s)q(s)e^{-\tau_d s}}{1 - q(s)e^{-\tau_d s}}. \tag{6.8}$$

Algebraic manipulation of (6.8) results in the equivalent form

$$C(s) = \frac{1 - (1 - b(s))q(s)e^{-\tau_d s}}{1 - q(s)e^{-\tau_d s}}. \tag{6.9}$$

The transfer function of the overall closed-loop system in Figure 6.1, which is also the complementary sensitivity transfer function $T(s)$, is obtained as

$$T(s) = \frac{C(s)G(s)}{1 + C(s)G(s)}. \tag{6.10}$$

Substituting for the repetitive controller's transfer function $C(s)$ from (6.9) into (6.10) results in

$$T(s) = \frac{\frac{1-(1-b(s))q(s)e^{-\tau_d s}}{1-q(s)e^{-\tau_d s}} G(s)}{1 + \frac{1-(1-b(s))q(s)e^{-\tau_d s}}{1-q(s)e^{-\tau_d s}} G(s)} \tag{6.11}$$

which after simplification becomes

$$T(s) = \frac{\left(1 - (1 - b(s))q(s)e^{-\tau_d s}\right)G(s)}{(1 + G(s)) - q(s)\left[1 + (1 - b(s))G(s)\right]e^{-\tau_d s}}. \tag{6.12}$$

Dividing the numerator and denominator of (6.12) by $1 + G(s)$ to obtain the closed-loop plant without the repetitive controller $C(s)$ results in

$$T(s) = \frac{\left(1 - (1 - b(s))q(s)e^{-\tau_d s}\right)\frac{G(s)}{1+G(s)}}{1 - q(s)\left[1 - b(s)\frac{G(s)}{1+G(s)}\right]e^{-\tau_d s}}. \tag{6.13}$$

The characteristic equation of the repetitive control system is thus

$$1 + \frac{Q(s)}{P(s)}e^{-\tau_d s} = 1 - q(s)\left(1 + (1 - b(s))\frac{G(s)}{1 + G(s)}\right)e^{-\tau_d s} = 0, \tag{6.14}$$

where the first expression on the left-hand side is the general equation for the characteristic polynomial of a time-delayed system.

Assume that $1 + G(s)$ has no zeros in the right half of the complex plane and τ_d satisfies (6.7). Then, noting the definition in (6.2), the regeneration spectrum $\mathscr{R}(\omega)$ for the plug-in repetitive system, shown in Figure 6.1, is

$$\mathscr{R}(\omega) = \left|q(j\omega)\left[1 - b(j\omega)\frac{G(j\omega)}{1 + G(j\omega)}\right]\right| < 1. \tag{6.15}$$

The expression for regeneration spectrum comprises two distinct filters – $q(s)$ and $b(s)$ – which have different effects on the stability of the repetitive control system.

If $1 + G(s) = 0$ has no roots in the right half plane, and also if $\mathscr{R}(\omega) < 1$ holds for all $\omega \in [0, \infty)$, then the repetitive control system is asymptotically stable. The first condition indicates that the closed-loop control system needs to be stable without the repetitive controller. The second condition simply shows that $q(s)$ should be chosen as a low-pass filter with unity d.c. gain. The expression within parentheses in (6.15) is approximately one as ω goes to infinity since $G(j\omega) \approx 0$ at high frequencies. The $q(s)$ filter should be lower than one at high frequencies to guarantee the stability. As clearly seen in (6.15), the use of the $b(s)$ filter can dramatically improve the relative stability. If the $b(s)$ filter is chosen to compensate amplitude and phase of $G(j\omega)/(1 + G(j\omega))$, then the magnitude of the term within the parenthesis in (6.15) will be close to zero for a broader range of frequencies.

It is worth noting that the sufficient condition (6.15) becomes a necessary and sufficient condition when τ_d gets large values. In those cases, the regeneration spectrum (6.15) also provides a good approximation of dominant roots of the characteristic equation (6.14). The assumption of large time-delays – in the sense of (6.7) – can be adequately practiced for many repetitive control applications; see, e.g., [31]. Equation (6.7) states the slowest time constant of the system in the absence of repetitive control should be larger than the time-delay τ_d. Before employing the controller design procedure described in the next section, one needs to choose the compensated plant transfer function $G(s)$ to satisfy (6.7).

Applying the Nyquist criterion to (6.14), one can determine standard relative stability measures, such as gain and phase margins. These margins describe the gain increase and additional phase lag due to the (possibly compensated) plant. The effect of altering parameters in the $q(j\omega)$ and $b(j\omega)$ filters on these measures is intricate to understand, therefore; relative stability measures mentioned above is not very beneficial in designing repetitive controllers. However, the regeneration spectrum provided in (6.15) clearly exhibits how the change of parameters in the $q(j\omega)$ filter affects the stability. Observing the effect of altering parameters in the $b(j\omega)$ filter on regeneration spectrum is moderately more challenging.

6.3   Repetitive controller basics

The working principle and elements of the repetitive control system, shown in Figure 6.1, are investigated in detail.

6.3.1   Internal model principle

The internal model principle, presented by Francis and Wonham [1], is in servo-system design for steady-state error characterisation. The general structure of the internal model system, extensively used in control of servo-systems, is illustrated in Figure 6.2. The internal model principle is explained below in detail with its mathematical theory and proof.

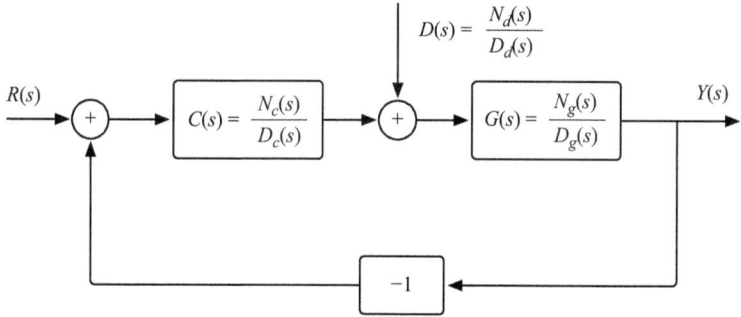

$$D(s) = \frac{N_d(s)}{D_d(s)}$$

Figure 6.2 Internal model control structure

Theorem 6.1. Internal model principle. *Assume that $N_g(s) = 0$ and $D_d(s) = 0$ do not have any common roots. If the closed-loop control system is asymptotically stable, and $D_c(s)$ can be factorised as $D_c(s) = D_d(s)D'_c(s)$, then the disturbance is asymptotically rejected.*

Proof: The steady-state error response to the disturbance $D(s)$ can be expressed as

$$\frac{E(s)}{D(s)} = -\frac{G(s)}{1 + C(s)G(s)} = -\frac{D_c(s)N_g(s)}{N_c(s)N_g(s) + D_c(s)D_g(s)}.$$

Since $D(s) = \frac{N_d(s)}{D_d(s)}$ and $D_c(s) = D_d(s)D'_c(s)$,

$$E(s) = -\frac{D_c(s)N_g(s)}{N_c(s)N_g(s) + D_c(s)D_g(s)}\frac{N_d(s)}{D_d(s)},$$

$$= -\frac{D'_c(s)N_g(s)N_d(s)}{N_c(s)N_g(s) + D_c(s)D_g(s)}.$$

Since all roots of $N_c(s)N_g(s) + D_c(s)D_g(s) = 0$ are located on the left-half plane, $e(t)$ converges to zero as t goes to infinity. ☐

Internal Model Principle (IMP) implies that a controlled output can track a class of reference commands or reject a class of disturbances without any steady-state error if a stable closed-loop control system contains the generator of this reference command (or disturbance). For instance, no steady-state error occurs for a step reference (or disturbance) signal in type 1 stable feedback system, which has a free integrator $1/s$ (i.e., the generator of the step function) in this loop.

An extension of the internal model principle for periodic references (or disturbances) is proposed by Hara *et al.* [29]. For periodic signals to achieve asymptotic tracking of a reference input (or rejection of a disturbance), the feedback loop needs to contain a generator of periodic signals, i.e., poles at the frequency of the periodic signal and its entire non-zero harmonics. This can be easily achieved by inclusion of the so-called periodic signal generator within the loop and is discussed next.

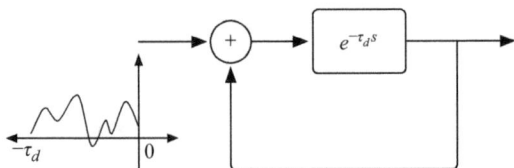

Figure 6.3   Periodic signal generator with an appropriate initial function

6.3.2   Periodic signal generator

A periodic signal with a known period can be generated by the signal generator structure, shown in Figure 6.3, with an appropriate initial function. Based on an extension of the Internal Model Principle (IMP) presented earlier, it is well-known that the controlled output can track a periodic reference command (or reject a periodic disturbance) without any steady-state error if the closed-loop control system includes the generator for this periodic command. This periodic signal generator includes a time-delay element, which is equal to the period of the reference (or disturbance), under a positive feedback loop. The transfer function of the signal generator, shown in Figure 6.3, is given by

$$G_{\text{psg}}(s) = \frac{e^{-\tau_d s}}{1 - e^{-\tau_d s}}. \tag{6.16}$$

The periodic signal generator, provided in (6.16), has infinitely many poles on the imaginary axis, i.e., $\mathbf{roots}\{1 - e^{-\tau_d s}\} \in \{\pm jk\omega : k \in \mathbb{N}_0, \omega = {}^{2\pi}/\tau_d\}$. The positive feedback loop in the repetitive controller generates the periodic signal needed at the plant input to asymptotically track the periodic reference signal accurately or reject the periodic disturbance effectively. The sensitivity transfer function of the closed-loop control system, shown in Figure 6.1, is given by

$$S(s) = \frac{1 - qe^{-\tau_s s}}{(1 + G)\left(1 - q(s)\left(1 - b(s)\frac{G(s)}{1 + G(s)}\right)e^{-\tau_s s}\right)}. \tag{6.17}$$

The closed-loop system, which includes a repetitive controller, has infinite loop gain at the fundamental frequency of $^1/\tau_d$ and its harmonics, and the sensitivity function of the closed-loop system, in turn, becomes zero at these frequencies. Although the infinite loop gain (or consequently zero sensitivity) provides perfect tracking and rejection capabilities at the frequencies mentioned above, it might cause instability at high frequencies. To tackle this stability issue, one should choose $q(s)$, seen in (6.17), as a low-pass filter with unity d.c. gain to attenuate the content above its cutoff frequency. The bandwidth of the $q(s)$ filter determines the frequency band of the periodic reference command to be tracked (or periodic disturbance to be rejected). There is, therefore, a trade-off between stability robustness and steady-state accuracy.

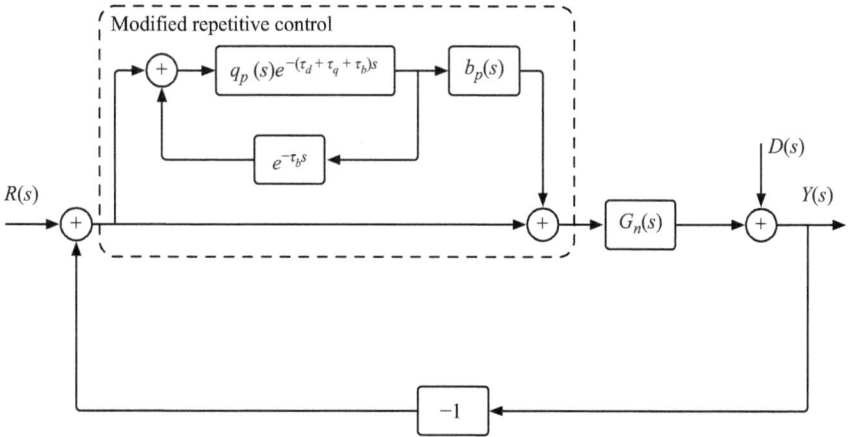

Figure 6.4 Modified repetitive control structure

6.3.3 Time advance

Elimination of phase lag in a closed-loop system can improve the tracking performance and also relative stability. One efficient way of cancelling phase lag is to introduce a time advance, which produces a phase shift that is linearly proportional to the frequency. For continuous-time systems, the time advance can be represented as $e^{\tau s}$. The main feature of the phase cancellation with time advance in repetitive control systems is to completely cancel the phase lag up to the cut-off frequency of the low-pass filter since the ability of the repetitive controller to reject periodic disturbances sharply deteriorates beyond this frequency. Bear in mind that the positive phase shift of the time advance does not have any adverse effect on the system [53].

Minimisation of $\mathscr{R}(\omega)$, provided in (6.15), enhances both the relative stability and tracking performance of a repetitive control system. The idea is to reduce that to being as close to zero as possible within the chosen bandwidth of the closed-loop (repetitive) control system. Recalling the form of (6.15), this will be achieved if $b(s)$ is used to reduce the phase of the closed-loop plant $G(s)/(1 + G(s))$ and if $q(s)$ is used to reduce the phase of $1 - b(s)G(s)/(1 + G(s))$. Cancellation of these phase shifts is possible as both $b(s)$ and $q(s)$ can be made to have phase advances as compared to the standard phase lag behaviour. This is possible in implementation and is illustrated in Figure 6.4 where the small phase advances in $b(s)$ and $q(s)$ are absorbed in the much larger phase lag of the repetitive control system. It is worth noting that the relative stability of repetitive control system is independent of the phase lag of the low-pass filter $q(s)$ but the tracking performance of the system will be adversely affected if this phase lag is not cancelled. Nevertheless, the phase shift of the dynamic compensator $b(s)$ is directly linked to the relative stability of repetitive control system; see, e.g., [53].

To guarantee the stability of the repetitive control system under the worst-case scenarios, the regeneration spectrum condition, i.e., $\|\mathscr{R}(\omega)\|_\infty < 1$ for all $\omega \in [0, \infty)$,

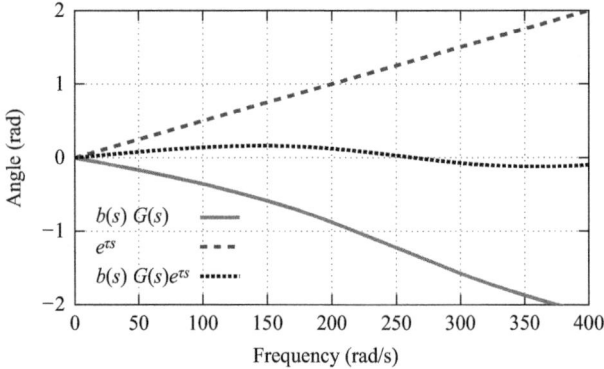

Figure 6.5   Compensation of the phase lag by the time advance

is not sufficient. Indeed, this condition should be modified as $\angle \mathscr{R}(\omega) \approx 0$ at all frequencies. It means that the phase shift of the regeneration spectrum should be approximately equal to zero. The best phase cancellation is the minimum excursion from $0°$ when the total phase $\angle \mathscr{R}(\omega)$ is added to the phase due to the time advances, such as τ_q and τ_b.

Example 6.1

The satellite mirror servo-control problem is used as a numerical example to demonstrate how to cancel the phase lag of a control system. To this end, the transfer function of the closed-loop system is given by

$$G(s) = \frac{439.8^2}{s^2 + 247.3s + 439.8^2}. \tag{6.18}$$

The dynamic compensator, selected as the approximate inverse of $G(s)$, is given by

$$b(s) = \frac{\frac{s^2}{439.8^2} + \frac{247.3s}{439.8^2} + 1}{\frac{s^2}{300^2} + \frac{s}{300} + 1}. \tag{6.19}$$

The time advance of the dynamic compensator is chosen as $\tau_b = 0.005$ s. As can be seen in Figure 6.5, the time advance $e^{\tau_b s}$ perfectly eliminates the phase lag of the closed-loop system within the chosen bandwidth.

6.3.4   Low-pass filter q(s) and dynamic compensator b(s)

Consider the repetitive control structure, shown in Figure 6.1, where G_n is the nominal model of the plant, Δ_m is the normalised unstructured multiplicative model uncertainty, W_T is the multiplicative uncertainty weighting function, and τ_d is the period of the periodic exogenous signal. The $q(s)$ and $b(s)$ filters are used for tuning the repetitive controller. Repetitive control systems can track periodic signals very accurately and can reject periodic disturbances very satisfactorily because the positive feedback loop, illustrated in Figure 6.1, is a generator for periodic signals with period τ_d when $q(s) = 1$. A low-pass filter with unity d.c. gain is used to improve robustness of stability [29,37].

The repetitive controller design involves the design of the two filters – $q(s)$ and $b(s)$ – as seen in Figure 6.1. The $q(j\omega)$-filter is ideally a low-pass filter which would be one in the frequency range of interest and zero at higher frequencies. It is, however, not possible to construct such a filter because $q(j\omega)$ will have negative phase angle, which will make $q(j\omega)$ differ from one, resulting in reduced accuracy. A small time advance is customarily incorporated into $q(s)$ to cancel out the negative phase of its low-pass filter part within its bandwidth. The time-delay, τ_d, corresponding to the period of the exogenous input signal, absorbs the time advance τ_q since one is much larger than other, i.e., $\tau_d \gg \tau_q$. Hence, this does not constitute an implementation problem; see Figure 6.4.

The main objective of the use of the dynamic compensator $b(s)$ is to improve the relative stability, the transition response and the steady-state accuracy in combination with the low-pass filter $q(s)$. Consider the function of frequency given by

$$\mathscr{R}(\omega) \triangleq \left| q(j\omega) \left[1 - b(j\omega) \frac{G(j\omega)}{1 + G(j\omega)} \right] \right|, \tag{6.20}$$

which is called the regeneration spectrum in [31]. $\mathscr{R}(\omega) < 1 - \epsilon$ for all $\omega \in [0, \infty)$ and some positive ϵ is a sufficient condition for stability [31]. Moreover, $\mathscr{R}(\omega)$ can be employed to obtain a good approximation of the locus of the dominant characteristic roots of the repetitive control system for large time delay, thus resulting in a measure of relative stability, as well. Accordingly, the compensator $b(s)$ is designed to approximately invert $G/(1 + G)$ within the bandwidth in an effort to minimise $\mathscr{R}(\omega)$. The dynamic compensator $b(s)$ can be selected as only a small time advance or time advance multiplied by a low-pass filter in order to minimise $\mathscr{R}(\omega)$. In order to make $\mathscr{R}(\omega) < 1$, the time advance in the filter $b(s)$ is chosen to cancel out the negative phase of $G/(1 + G)$. This small time advance can easily be absorbed by the much larger time delay τ_d corresponding to the period of the exogenous input signal and does not constitute an implementation problem; see Figure 6.4.

The $q(s)$ and $b(s)$ filters are thus expressed as

$$q(s) = q_p(s)e^{\tau_q s}, \tag{6.21}$$

and

$$b(s) = b_p(s)e^{\tau_b s}. \tag{6.22}$$

The time advances τ_q and τ_b are firstly chosen to decrease the magnitude of $\mathscr{R}(\omega)$ given in (6.20). Then, the design focuses on pairs of chosen parameters in $q_p(s)$ or $b_p(s)$ to satisfy a frequency domain bound on the robust performance criterion.

If $L(s)$ denotes the loop gain of a control system, its sensitivity and complementary sensitivity transfer functions are

$$S(s) \triangleq \frac{1}{1 + L(s)}, \tag{6.23}$$

and

$$T(s) \triangleq \frac{L(s)}{1 + L(s)}. \tag{6.24}$$

The parameter space design, presented in the next section, aims at satisfying the condition:

$$|W_S S| + |W_T T| < 1, \qquad \forall \omega \in [0, \infty). \tag{6.25}$$

This condition is similar to satisfying the robust performance requirement $\| \, |W_S S| + |W_T T| \, \|_\infty < 1$ where W_S and W_T are sensitivity and complementary sensitivity function weights.

The loop gain of the repetitive control system, shown in Figures 6.1 and 6.4, is given by

$$L = G_n \left(1 + \frac{q_p}{1 - q_p e^{(-\tau_d + \tau_q)s}} b_p e^{(-\tau_d + \tau_q + \tau_b)s} \right). \tag{6.26}$$

The robust performance design requires

$$
\begin{aligned}
&|W_S(\omega) S(j\omega)| + |W_T(\omega) T(j\omega)| \\
&= \left| \frac{W_S(\omega)}{1 + L(j\omega)} \right| + \left| \frac{W_T(\omega) L(j\omega)}{1 + L(j\omega)} \right| < 1, \quad \forall \omega \in [0, \infty),
\end{aligned} \tag{6.27}
$$

or equivalently,

$$|W_S(\omega)| + |W_T(\omega) L(j\omega)| < |1 + L(j\omega)|, \quad \forall \omega \in [0, \infty), \tag{6.28}$$

to be satisfied.

6.4 Parameter space approach to repetitive control

Conventional controller design methodologies such as pole placement and trial-and-error approach give unique controllers for each plant. No further robustness criteria can be incorporated in the design step [54]. The parameter space approach can be used to determine a set of coefficients for a given controller structure that simultaneously stabilises a family of plants [54]. The parameter space approach allows determining the set of controller parameters, for which the characteristic polynomial is stable. The parameter vector might consist of uncertain plant parameters, or it might consist of coefficients of a fixed structure controller. Additionally, the parameter space approach can provide information about effects of parameter variations on stability and it can provide information about the closeness to instability upon parameter changes. The resulting controllers are not fragile because they are chosen in the inner part of the parameter space.

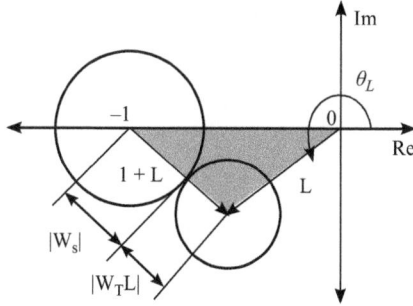

Figure 6.6   Illustration of the point condition for robust performance

6.4.1   Mapping robust performance frequency domain specifications into controller parameter space

In the present section, a repetitive controller design procedure based on mapping the robust performance frequency domain performance specification given in (6.28) with an equality sign into the chosen repetitive controller parameter plane at a chosen frequency is described.

Consider the robust performance problem given in Figure 6.6 illustrating (6.28) with an equality sign (called the robust performance point condition). Apply the cosine rule to the triangle with vertices at the origin, -1 and L in Figure 6.6 to obtain

$$(|W_S(\omega)| + |W_T(\omega)L(j\omega)|)^2 = |L(j\omega)|^2 + 1^2 + 2|L(j\omega)|\cos\theta_L. \tag{6.29}$$

Equation (6.29) is a quadratic equation in $|L(j\omega)|$ and its solutions are

$$|L(j\omega)| = \frac{(-\cos\theta_L + |W_S(\omega)||W_T(\omega)|) \pm \sqrt{\Delta_M}}{1 - |W_T(\omega)|^2}, \tag{6.30}$$

where

$$\Delta_M = \cos^2\theta_L + |W_S(\omega)|^2 + |W_T(\omega)|^2 - 2|W_S(\omega)|^2|W_T(\omega)|^2\cos\theta_L - 1. \tag{6.31}$$

Only, positive and real solutions for $|L(j\omega)|$ are allowed, i.e., $\Delta_M \geq 0$ in (6.30) must be satisfied. The point condition solution procedure is outlined below.

M1   Define the set of frequencies to be used as

$$\Omega \triangleq \left\{ \omega_k = \frac{2\pi k}{\tau_d}, \ k \in \mathbb{N}_0 : \omega_1, \ldots, \omega_n; \omega_{n+1}, \ldots, \omega_m; \omega_{m+1}, \ldots, \omega_l \right\},$$

where $\omega_1 = 2\pi/\tau_d$ is the frequency of the periodic exogenous input and $\omega_k = 2\pi k/\tau_d$ is the chosen bandwidth of repetitive control (limited by the bandwidth of the actuator used). Frequencies ω_{m+1} to ω_l are high frequencies where significant model uncertainty exists ($\omega_{m+1} > 10\omega_n$) and the intermediate frequencies ω_{n+1} to ω_m.

Remark 6.1. *It is worth noting that the inherent time delay in a repetitive control system will improve performance only at the fundamental frequency $\omega_1 = 2\pi/\tau_d$ and its harmonics. Repetitive control will, however, worsen performance at frequencies between the fundamental frequency and its harmonics. For this reason, repetitive control is only used in the presence of an external periodic input (reference or disturbance) as it will result in degraded performance for non-periodic external inputs. For that reason, the weights ($W_S(\omega)$ and $W_T(\omega)$) can be assumed to be zero outside the finite set Ω given in step M1. Once the design is complete, the designer checks the $|S(j\omega)|$ plot at low frequencies and the $|T(j\omega)|$ plot at high frequencies to make sure that the magnitude envelopes corresponding to intermediate frequencies (between the harmonics) are at an acceptable level. Another approach will be to specify weights for intermediate frequencies in between the fundamental frequency and harmonics. A significant deficiency of parameter space methods is that one needs to use a large number of frequencies in the set Ω to make sure that the robust performance condition will not be violated at frequencies outside of Ω. In the case of repetitive control, this problem is less severe at low frequencies where the designer is interested mainly in reducing the sensitivity function at the fundamental frequency and its harmonics. A large number of frequency points can be used at higher frequencies above the bandwidth of the repetitive control system.*

M2 Choose a specific frequency value $\omega - \omega_i \in \Omega$ for any $i \in \{1, 2, \ldots, l\}$ from set Ω in step M1. $|W_S(\omega)|$, $|W_T(\omega)|$ and $|G(j\omega)|$ at a frequency ω are known at this point.

M3 Let $\theta_L \in [0, 2\pi]$. Evaluate Δ_M by using (6.31) and select the active range of θ_L where $\Delta_M \geq 0$ is satisfied. For all values of θ_L in the active range:

    M3a Evaluate $|L(j\omega)|$ by using (6.30). Keep only the positive solutions.

    M3b Evaluate $L(j\omega) = |L(j\omega)|e^{j\theta_L}$.

    M3c Solve for the corresponding repetitive controller filters $q_p(j\omega)$ and $b_p(j\omega)$ at the chosen frequency ω by utilising

$$q_p(j\omega) = \frac{L(j\omega) - G(j\omega)}{L(j\omega) - G(j\omega)[1 - b(j\omega)]}e^{(\tau_d - \tau_q)j\omega} \tag{6.32}$$

and

$$b_p(j\omega) = [L(j\omega) - G(j\omega)]\left[\frac{1 - q(j\omega)e^{-\tau_d j\omega}}{q(j\omega)e^{-\tau_d j\omega}}\right]e^{-\tau_b j\omega}. \tag{6.33}$$

    M3d Using the specific structure of $q_p(j\omega)$ or $b_p(j\omega)$, back solve for the two chosen controller parameters within them. For instance, $q_p(s)$ and $b_p(s)$ can be chosen as a multiplication of the second order controllers given by

$$q_p(s) = \prod_{i=1}^{n} \frac{q_{5i}s^2 + q_{4i}s + q_{3i}}{q_{2i}s^2 + q_{1i}s + q_{0i}}, \tag{6.34}$$

Table 6.1   Controller coefficients table

Control action	n	q_{5i}	q_{4i}	q_{3i}	q_{2i}	q_{1i}	q_{0i}
P	1	0	0	K	0	0	1
PD	1	0	KT_d	K	0	0	1
PI	1	0	K	KT_i	0	1	0
PID	1	KT_d	K	KT_i	0	1	0
Lag ($\beta > 1$)	1	0	KT	K	0	βT	1
Lead ($0 < \alpha < 1$)	1	0	KT	K	0	αT	1
1st Order filter	1	0	0	K	0	τ	1
2nd Order filter	1	0	0	$K\omega^2$	1	$2\zeta\omega$	ω^2

and

$$b_p(s) = \prod_{i=1}^{n} \frac{b_{5i}s^2 + b_{4i}s + b_{3i}}{b_{2i}s^2 + b_{1i}s + b_{0i}}. \tag{6.35}$$

There are six tuneable parameters for $n = 1$ in (6.35) which can be used to represent different types of controllers. These six tuneable parameters are q_{5i}, q_{4i}, q_{3i}, q_{2i}, q_{1i} and q_{0i} for $q_p(s)$. For $n = 1$, the controller structure in (6.35) consists of some well-known controller types such as proportional-integral-derivative (PID), lead-lag controller, first or second order filters, as illustrated in Table 6.1. If the performance of the filters which are utilised in the repetitive controller for $n = 1$ is unsatisfactory, n can be increased and new higher order filters can be synthesised. For the filter structure choice in (6.35), the back solution procedure uses

$$\begin{aligned}
\text{Re}[q_p(j\omega)] &= \frac{(q_{3i} - q_{5i}\omega^2)(q_{0i} - q_{2i}\omega^2) + q_{1i}q_{4i}\omega^2}{(q_{0i} - q_{2i}\omega^2)^2 + (q_{1i}\omega)^2}, \\
\text{Im}[q_p(j\omega)] &= \frac{q_{4i}\omega(q_{0i} - q_{2i}\omega^2) - q_{1i}\omega(q_{3i} - q_{5i}\omega^2)}{(q_{0i} - q_{2i}\omega^2)^2 + (q_{1i}\omega)^2},
\end{aligned} \tag{6.36}$$

and

$$\begin{aligned}
\text{Re}[b_p(j\omega)] &= \frac{(b_{3i} - b_{5i}\omega^2)(b_{0i} - b_{2i}\omega^2) + b_{1i}b_{4i}\omega^2}{(b_{0i} - b_{2i}\omega^2)^2 + (b_{1i}\omega)^2}, \\
\text{Im}[b_p(j\omega)] &= \frac{b_{4i}\omega(b_{0i} - b_{2i}\omega^2) - b_{1i}\omega(b_{3i} - b_{5i}\omega^2)}{(b_{0i} - b_{2i}\omega^2)^2 + (b_{1i}\omega)^2}.
\end{aligned} \tag{6.37}$$

M4   The solution in step M3 above results in a closed curve which is plotted for solving 2 of the 12 parameters q_{0i} to q_{5i} and b_{0i} to b_{5i}. Plot the closed curve obtained in the chosen controller parameter space. Either the inside (drawn with a solid boundary) or outside (drawn with a dashed boundary) of this curve is a solution of (6.27) at the chosen frequency (see, e.g., the ellipses in Figure 5). The region obtained is the point condition solution in the chosen repetitive controller parameter plane at the frequency chosen in step M2.

M5   Go back to step M2 and repeat the procedure at a different frequency until all frequencies in set Ω are used.

M6   Plot the intersection of all point condition solutions for all frequencies in set Ω. This is the overall solution region for the robust performance requirement.

As the solution procedure only uses frequency response data and is numerical in nature, plants with time delay or poles on the imaginary axis and discontinuous weights do not pose any problems. Note that solution regions for nominal performance $|W_S(\omega)S(j\omega)| < 1$ for all $\omega \in \Omega$ and for robust stability $|W_T(\omega)T(j\omega)| < 1$ for all $\omega \in \Omega$ can easily be obtained using the algorithm above by setting $W_S = 0$ and $W_T = 0$, respectively. It is then possible to concentrate on nominal performance at low frequencies, robust performance at intermediate frequencies and robust stability at high frequencies, obtaining three solution regions. The overall solution region in the controller parameter space is then determined by the intersection of all three regions for nominal performance, robust performance and robust stability. This procedure is easily programmable and quickly results in a controller parameter space representation of the solution. The controller parameter space presentation obtained offers the ease of visualisation of parameter space methods (see, e.g., Figure 6.8) when one accepts the shortcoming of treating only two controller parameters at a time. Multi-objective design can easily be formed in parameter space as it amounts simply to the intersection of individual solution regions. It is also possible to determine the final design (or tuning point) by optimising some other criteria, such as nominal time domain performance within the solution region obtained. In contrast to $\mathcal{H}_\infty$ optimal control synthesis, there is no relationship between the order of repetitive control filters and the complexity of weights in this proposed method. The main strength of this approach is that low-order, easily implementable repetitive control filters are specified from the beginning.

It is possible that for certain data sets $|W_S|$, $|W_T|$, G, ω; no solutions to the solution procedure outlined above exist. Non-existence of a solution for a specific frequency ω could be because of non-existence of a positive Δ_M in (6.31) or non-existence of a positive solution $|L|$ in (6.30). Non-existence of a solution usually results from a weight $|W_S|$ or $|W_T|$ that is too restrictive. The solution procedure, which is programmed in an interactive fashion, results in no solution points in this case. Then, the user will know that his robust performance requirement at that frequency was too restrictive and has the choice of relaxing this requirement. Note that solutions might exist at all individual frequencies, however; their intersection in Step M6 resulting in the overall solution region, might still be empty. In that case, the user must change the sensitivity and complementary sensitivity weights at the problematic frequencies.

6.5   Case study: high-speed atomic force microscope scanner position control

In this part of Chapter 6, the high-speed AFM scanner, which is designed and modelled in [55], is used as a numerical example to explain the methodology of the multi-objective parameter space approach for SISO repetitive controller design. The second-order and fourth-order mathematical models of this high-speed AFM scanner are given in [55]. In this example, the fourth-order model is used since it includes the first mode of the piezoelectric stack in the vertical direction. The transfer function of the AFM scanner is given by

$$G(s) = \frac{K(s^2 + 2\zeta_2\omega_2 s + \omega_2^2)}{(s^2 + 2\zeta_1\omega_1 s + \omega_1^2)(s^2 + 2\zeta_3\omega_3 s + \omega_3^2)}, \tag{6.38}$$

where $K = 1 \times 10^{12}$ nm/V includes the power amplifier and sensor gain. The system seen in (6.38) has two resonant frequencies and one anti-resonant frequency. The numerical values of these frequencies are given as $f_1 = 40.9$ kHz, $f_2 = 41.6$ kHz and $f_3 = 120$ kHz and can be seen in Figure 6.7. The numerical values of the relative damping coefficients are given as $\zeta_1 = 0.016$, $\zeta_2 = 0.016$ and $\zeta_3 = 0.17$.

The dynamic compensator $b(s)$ is chosen as a pure time advance as

$$b(s) = b_p(s)e^{\tau_b} = e^{3 \times 10^{-6} s}. \tag{6.39}$$

The low-pass filter $q(s)$ is chosen as

$$q(s) = q_p(s)e^{\tau_q s} = \frac{a_0}{s^2 + a_1 s + a_0}e^{7.5 \times 10^{-6} s}. \tag{6.40}$$

The parameters of $q_p(s)$ given in (6.40) are chosen as $q_{51} = q_{41} = 0$, $q_{21} = 1$, $q_{31} = q_{01} = a_0$ and $q_{11} = a_1$ in the general form (6.35) to obtain unity d.c. gain. Phase advance is also added to this low-pass filter phase cancellation. Thus, a decrease in the steady-state error is aimed. The region in the $a_0 - a_1$ controller parameter space are computed for three cases which are respectively the nominal performance at low frequencies ($W_T = 0$), robust performance at intermediate frequencies and robust stability at high frequencies ($W_S = 0$).

The sensitivity constraints are specified at a set of discrete frequencies. The periodic input command of the high-speed AFM scanner has a period of τ_d s. The specific numerical values of the chosen weights used in the computation of the controller parameters are seen in Table 6.2. The frequencies corresponding to the weights in Table 6.2 are shown with dots in the Bode magnitude plot of Figure 6.7. The

Table 6.2   Desired sensitivity magnitude upper bounds at $\tau_d = 0.0005$ s

Frequency Range	k	$f = k/\tau_d (kHz)$	W_S	W_T
Low (NP)	1	2	500	0
Low (NP)	2	4	225	0
Low (NP)	3	6	115	0
Low (NP)	4	8	75	0
Intermediate (RP)	40	80	3.3	0.001
Intermediate (RP)	50	100	4.5	0.045
Intermediate (RP)	55	110	4.5	0.001
Intermediate (RP)	60	120	1.5	0.005
Intermediate (RP)	70	140	1.5	0.01
High (RS)	80	160	0	0.05
High (RS)	90	180	0	0.05
High (RS)	100	200	0	0.05

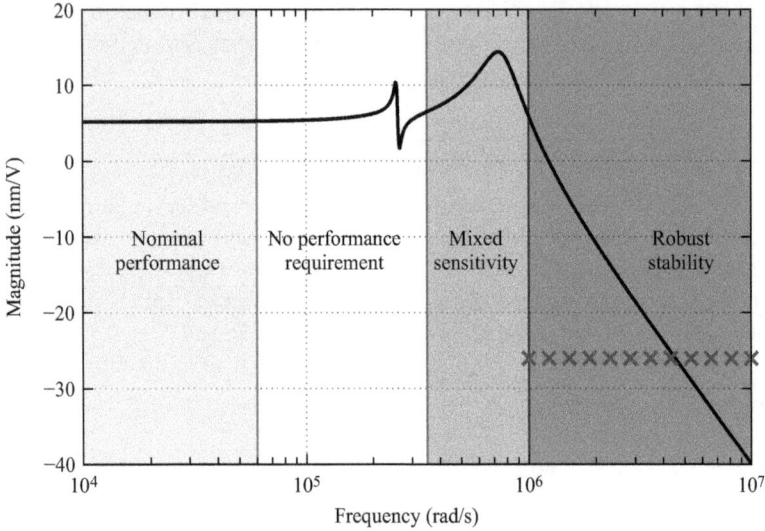

Figure 6.7   The Bode magnitude plot of high-speed AFM-scanner with the mapping frequencies for the nominal performance, robust performance and robust stability

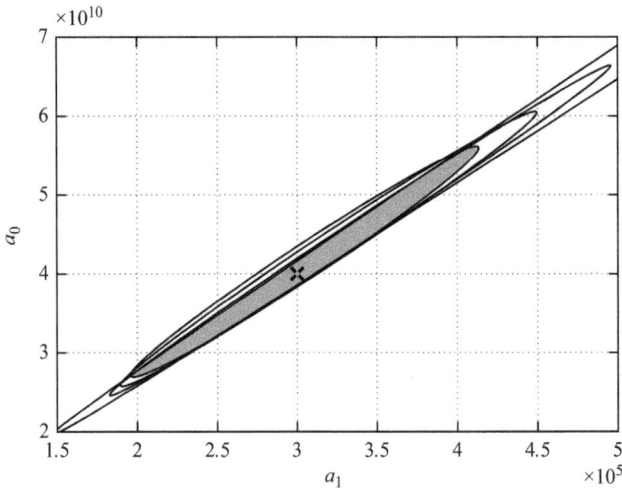

Figure 6.8   Parameter space for $|W_S(\omega)S(j\omega)| < 1$ and $|W_T(\omega)T(j\omega)| < 1$, $\forall \omega \in \Omega$

intersection of the solution regions for the nominal performance and stability robustness is presented in Figure 6.8, whereas the overall region calculated for nominal performance, robust performance and robust stability can be demonstrated in Figure 6.9. Note that the sufficient stability condition $\mathcal{R}(\omega) < 1$ for all $\omega \in \Omega$ is also

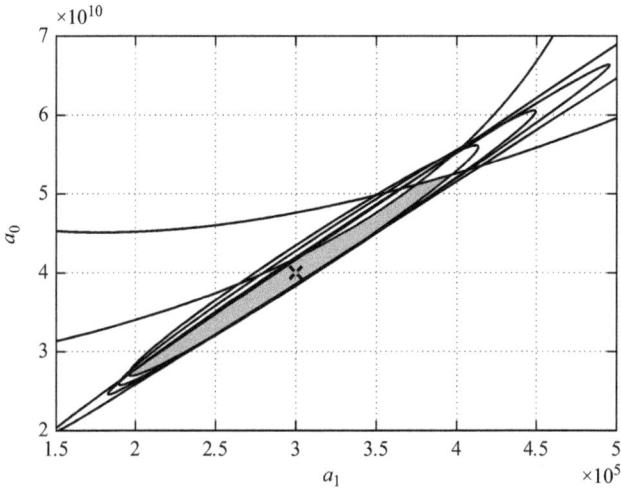

Figure 6.9 Parameter space for $|W_S(\omega)S(j\omega)| + |W_T(\omega)T(j\omega)| < 1, \forall \omega \in \Omega$

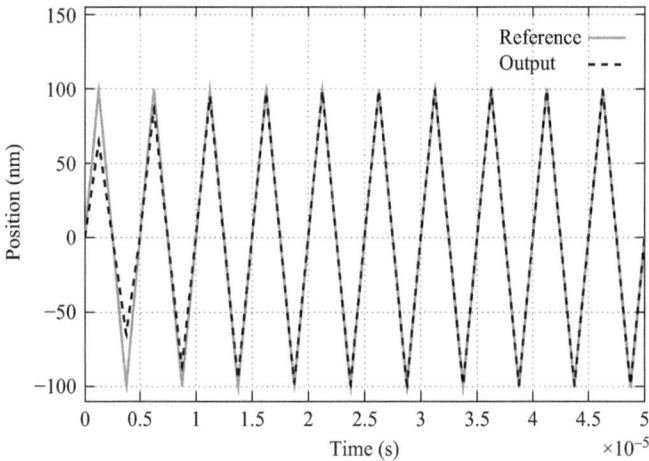

Figure 6.10 Output signal of the controlled system for triangular wave input at 2 kHz

mapped and nominal stability is thus satisfied for the two solution regions shown in Figure 6.9.

The mathematical model of the high-speed AFM scanner cannot be fitted very well for frequencies above 160 kHz. A uniform weight for the robust stability requirement for frequencies above this value is chosen here as $|T(j2\pi f)| < 0.05$ for all $f \geq 160$ kHz. This corresponding discontinuous weight W_T has been shown

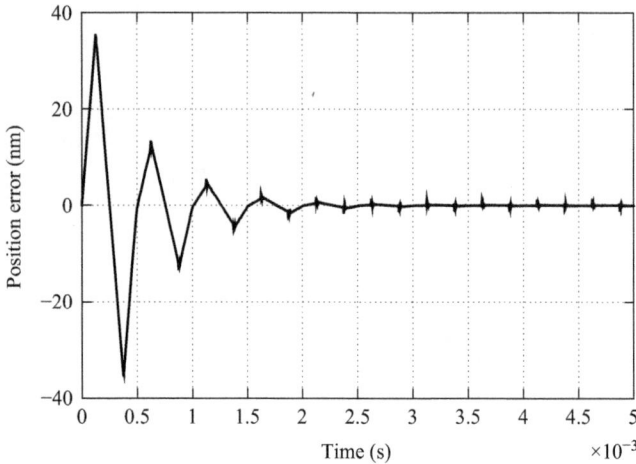

Figure 6.11   Error signal of the controlled system for triangular wave input at 2 kHz

graphically with gray-coloured cross sign in Figure 6.7. The relative multiplicative error $|(G - G_n)/G_n|$ has to be below the weight specified in the stability robustness considerations given in Figure 6.7. The intersection of the regions, which are calculated in order for the nominal performance, the robust performance and the robust stability requirements, in the $a_0 - a_1$ controller parameter space is filled with gray colour. The designation procedure is concluded by choosing a point in the controller parameter plane given in Figure 6.9. The solution within this region is chosen arbitrarily in this example and is point is marked with a cross in Figure 6.9. The simulation result for a triangular wave input with the period 2 kHz and amplitude can be seen in Figures 6.10 and 6.11. This result shows the effectiveness of the repetitive controller in decreasing the steady-state error while tracking a periodic input signal.

6.6   Case study: Quanser QUBE™ Servo system

We will re-visit the Quanser QUBE™ servo rotational speed control problem that was considered in Section 3.6. The plant model from control input to motor rotational position is given by

$$G(s) = \frac{28}{s(0.1s + 1)},$$
(6.41)

where the numerical values used were rounded to the nearest digit. The transfer function of the closed-loop control system with unity feedback is obtained as

$$G_{cl}(s) = \frac{G(s)}{1 + G(s)} = \frac{280}{s^2 + 10s + 280}.$$
(6.42)

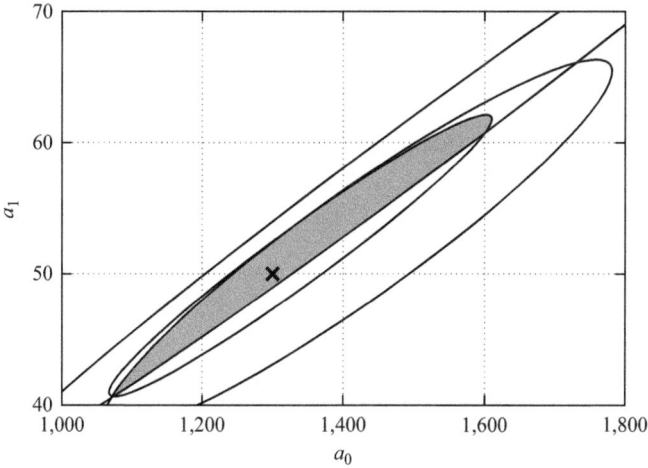

Figure 6.12 Parameter space for $|W_S(\omega)S(j\omega)| < 1$, $\forall \omega \in \Omega$

The dynamic compensator $b(s)$, which is equal to the approximate inverse of $G/(1+G)$ within the bandwidth of the low-pass filter $q(s)$, can be chosen as

$$b_p(s) = \frac{\frac{10}{7}s^2 + \frac{100}{7}s + 400}{s^2 + 40s + 400}. \tag{6.43}$$

To eliminate the phase shift of the regeneration spectrum (6.15), the time advance for the dynamic compensator $b(s)$ is chosen as

$$b(s) = b_p(s)e^{0.1s}. \tag{6.44}$$

The low-pass filter $q(s)$ with time advance is selected as

$$q(s) = q_p(s)e^{\tau_q s} = \frac{a_0}{s^2 + a_1 s + a_0}e^{0.04s}. \tag{6.45}$$

Notice that the parameters of $q_p(s)$ given in (6.45) are chosen as $q_{51} = q_{41} = 0$, $q_{21} = 1$, $q_{31} = q_{01} = a_0$ and $q_{11} = a_1$ in the general form (6.35) to obtain unity d.c. gain. In this example, the region in the $a_0 - a_1$ controller parameter space is computed for only the nominal performance at low frequencies (i.e., $W_T = 0$). The sensitivity constraints W_S, therefore, are chosen as 250, 25 and 4 at discrete frequencies 1 Hz, 2 Hz and 3 Hz, respectively. The reference command is chosen as a sinusoidal signal with a period $\tau_d = 1$ s. For different frequencies and corresponding sensitivity values, the solution procedure, described earlier, is applied, and the ellipsoids, seen in Figure 6.12, are computed. The intersection of these regions in the $a_0 - a_1$ parameter space is filled with gray colour. Lastly, the design procedure is concluded by arbitrarily choosing a point (marked with a cross) in the controller parameter space shown in Figure 6.12. As seen in Figure 6.13, the use of the repetitive controller reduces the steady-state error dramatically while tracking a periodic reference trajectory.

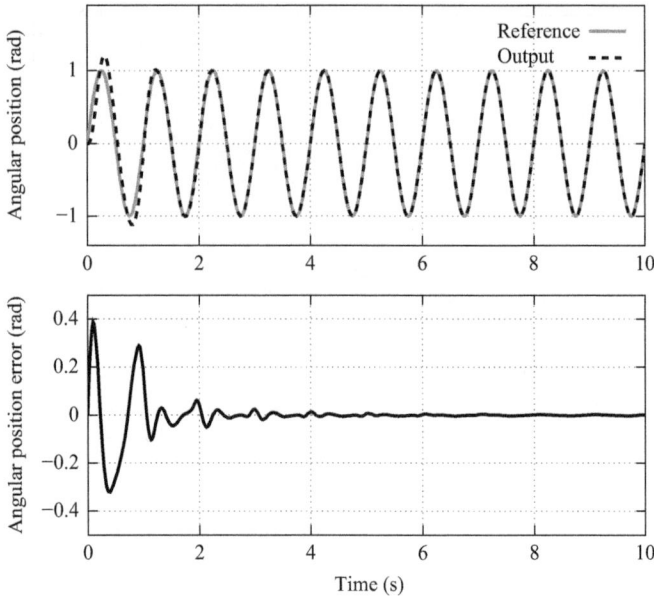

Figure 6.13   Simulation results for sinusoidal wave input at 1 Hz

6.7   COMES toolbox: repetitive control system design

In this section, we present a MATLAB®-based toolbox, seen in Figure 6.14, that implements parametric robust control methods for repetitive control design. The design technique is based on mapping a frequency domain mixed sensitivity bound into the chosen repetitive controller parameter plane. The solution procedure leads to graphical solution regions (2-D plots with colour filling showing where the specifications are met) in the controller parameter space. Our main goal is to design a user-friendly toolbox with graphical user interface (GUI), which hides all calculations from the user as much as possible. The user can thus focus on analysing the graphical results rather than doing all the complicated and tedious calculations. This MATLAB-based toolbox has two built-in examples: servo-hydraulic material testing machine position control and high-speed AFM scanner position tracking system.

6.8   Discrete-time repetitive control

Since most control systems are implemented in software on a dedicated computer, this section extends the parameter space design procedure, described in Section 6.4, to synthesise discrete-time repetitive controllers that satisfy a robust performance requirement. The discrete-time repetitive control system proposed here consists of two functional filters: a low-pass filter and a dynamic compensator. As mentioned

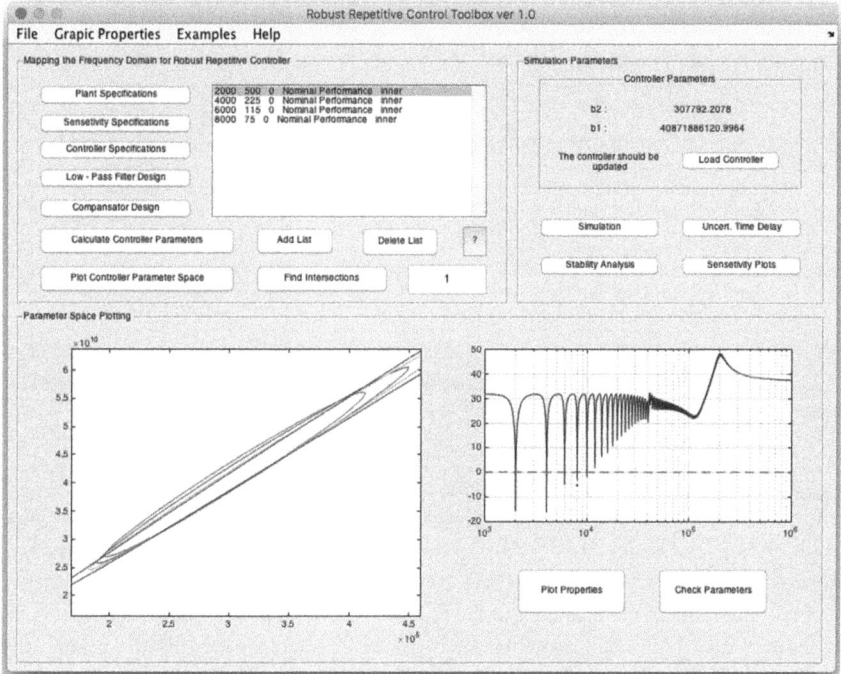

Figure 6.14   Main window of repetitive control system design part of COMES toolbox

earlier, the dynamic compensator $b(z)$ should be selected as the approximate inverse of $G/(1 + G)$ to improve the relative stability (by minimising the regeneration spectrum function). Suitable approximate inverse filter design techniques can be found in Chapter 4.

6.8.1   Basic elements of discrete-time repetitive control

We consider the repetitive control structure, depicted in Figure 6.15, where $G_n(z)$ is the nominal model of the plant, Δ_m is the normalised unstructured multiplicative model uncertainty, W_T is the multiplicative uncertainty weighting function, NT_s is the period of the periodic exogenous signal and T_s is the sampling time. The closed-loop system without repetitive controller, i.e., $G_n(z)/(1 + G_n(z))$, is stable. We use the low-pass filter $q(z)$ and the dynamic compensator $b(z)$ to tune the repetitive controller. Repetitive control systems can track periodic signals very accurately while rejecting periodic disturbances very satisfactorily since the positive feedback loop in Figure 6.15 is a generator of periodic signals with period NT_s for $q(z) = 1$; cf. [56]. Here, $q(z)$ is chosen as a low-pass filter with unity d.c. gain due to the robustness of the stability [29,37].

The design of discrete-time repetitive controllers involve the design of two filters $q(z)$ and $b(z)$ in Figure 6.15. In the frequency domain, the ideal low-pass filter $q(e^{j\omega T_s})$ would be unity within the frequency range of interest and zero at higher frequencies.

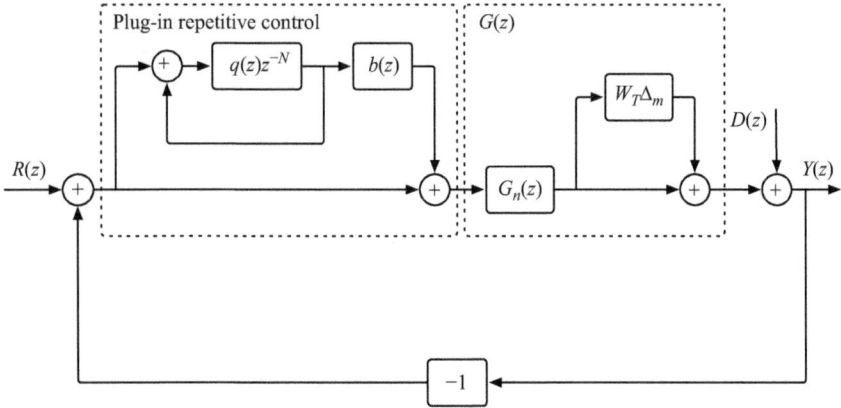

Figure 6.15   Plug-in (discrete-time) repetitive control structure

This is not possible in practice, furthermore; $q(e^{j\omega T_s})$ will typically have negative phase angle which will make $q(e^{j\omega T_s})$ differ from unity, resulting in reduced accuracy. To improve the accuracy of the repetitive controller, a small time advance is customarily incorporated into $q(z)$ to cancel out the negative phase of its low-pass filter part within its bandwidth. This small time advance can easily be absorbed by the much larger time delay NT_s corresponding to the period of the exogenous input signal and does not constitute an implementation problem as seen in Figure 6.16 where a small time advance in $q(z)$ has also been assumed.

The key objective of the usage of the dynamic compensator $b(z)$ is to improve the relative stability, the transition response and the steady-state accuracy in combination with the low-pass filter $q(z)$. Consider the function of frequency given by

$$\mathcal{R}(\omega) = \left| q(z) \left[1 - b(z) \underbrace{\frac{G_n(z)}{1 + G_n(z)}}_{G_{cl}(z)} \right] \right|_{z=e^{j\omega T_s}}$$

$$\triangleq |q(z)[1 - b(z)G_{cl}(z)]|_{z=e^{j\omega T_s}}, \qquad (6.46)$$

which is called the regeneration spectrum in [14]. The condition that $\mathcal{R}(\omega) < 1 - \epsilon$ for all frequencies and some positive ϵ is sufficient for stability [14]. Furthermore, $\mathcal{R}(\omega)$ can be utilised to obtain a good approximation of the locus of the dominant characteristic roots of the repetitive control system for large time delay (it is described as a sufficient and necessary condition in [14]), thus resulting in a measure of relative stability, as well. Accordingly, the compensator $b(z)$ is designed to approximately invert $G_n/(1 + G_n)$ in order to minimise $\mathcal{R}(\omega)$. The design of the approximate inverse of $G(z)/(1 + G(z))$ is explained in detail in the following section. In order to make $\mathcal{R}(\omega) < 1$, the time advance in the filter $b(z)$ is chosen to cancel out the negative phase of $G_n/(1 + G_n)$. This small time advance can easily be absorbed by the much larger time delay NT_s corresponding to the period of the exogenous input signal and does not constitute an implementation problem; see Figure 6.16.

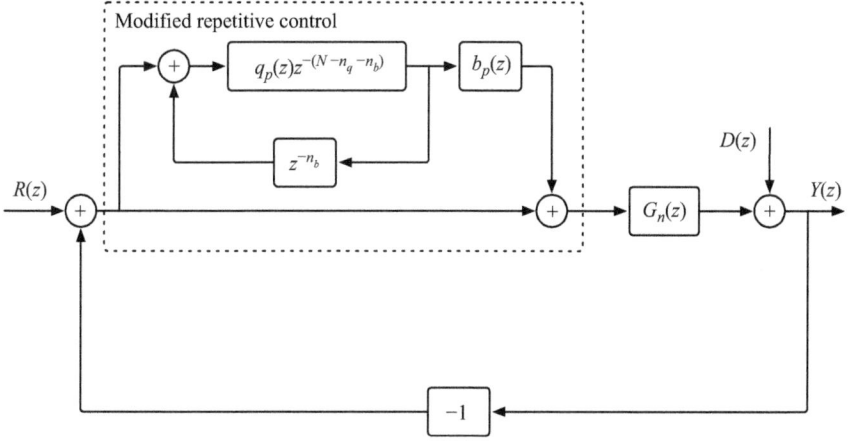

Figure 6.16 Modified (discrete-time) repetitive control structure

The $q(z)$ and $b(z)$ filters are thus expressed as $q(z) = q_p(z)z^{n_q}$ and $b(z) = b_p(z)z^{n_b}$. The time advances z^{n_q} and z^{n_b} are firstly chosen to decrease the magnitude of $\mathscr{R}(\omega)$ given in (6.46). Then, the design focuses on pairs of chosen parameters in $q_p(z)$ to satisfy a frequency domain bound on the robust performance criterion. If $L(z)$ denotes the loop gain of a control system, its sensitivity and complementary sensitivity transfer functions are written as

$$S(z) \triangleq \frac{1}{1 + L(z)} \quad \text{and} \quad T(z) \triangleq \frac{L(z)}{1 + L(z)}.$$

The parameter space design, presented in the following section, aims at satisfying the condition

$$|W_S(\omega)S(e^{j\omega T_s})| + |W_T(\omega)T(e^{j\omega T_s})| < 1, \quad \forall \omega \in [0, \infty), \tag{6.47}$$

which is similar to satisfying the robust performance requirement

$$\||W_S(\omega)S(e^{j\omega T_s})| + |W_T(\omega)T(e^{j\omega T_s})|\|_\infty < 1, \tag{6.48}$$

where $W_S(\omega)$ and $W_T(\omega)$ are sensitivity and complementary sensitivity function weights, respectively. The loop gain of the nominal repetitive control system, shown in Figures 6.15 and 6.16, is given by

$$L(z) = G_n(z)\left(1 + \frac{q_p(z)}{1 - q_p(z)z^{-N+n_q}}b_p(z)z^{-N+n_q+n_b}\right). \tag{6.49}$$

The robust performance design requires

$$
\begin{aligned}
&|W_S(\omega)S(e^{j\omega T_s})| + |W_T(\omega)T(e^{j\omega T_s})| \\
&= \left|\frac{W_S(\omega)}{1 + L(e^{j\omega T_s})}\right| + \left|\frac{W_T(\omega)L(e^{j\omega T_s})}{1 + L(e^{j\omega T_s})}\right| < 1,
\end{aligned}
\tag{6.50}
$$

or equivalently

$$|W_S(\omega)| + |W_T(\omega)L(e^{j\omega T_s})| < |1 + L(e^{j\omega T_s})|, \quad \forall \omega \in [0, \infty) \qquad (6.51)$$

to be satisfied.

6.8.2   Discrete-time repetitive control design procedure based on the parameter space approach

In this section, a discrete-time controller design procedure, which is based on mapping robust performance specifications, is presented. The robust performance problem is graphically illustrated in Figure 6.6. Apply the cosine rule to the shaded triangle in Figure 6.6 to obtain

$$\left(|W_S(\omega)| + |W_T(\omega)L(e^{j\omega T_s})|\right)^2 = |L(e^{j\omega T_s})|^2 + 1^2 + 2|L(e^{j\omega T_s})| \cos \alpha . \qquad (6.52)$$

Equation (6.52) is a quadratic equation in terms of $|L(e^{j\omega T_s})|$ and its solutions are given by

$$|L(e^{j\omega T_s})| = \frac{(-\cos \alpha + |W_S(\omega)||W_T(\omega)|) \pm \sqrt{\Delta_M}}{1 - |W_T(\omega)|}, \qquad (6.53)$$

where

$$\Delta_M = \cos^2 \alpha + |W_S(\omega)|^2 + |W_T(\omega)|^2 - 2|W_S(\omega)|^2|W_T(\omega)|^2 \cos \alpha - 1 . \quad (6.54)$$

Only positive and real solutions for $|L|$ are allowed, i.e., $\Delta_M \geq 0$ in (6.54) must be satisfied. The point condition solution procedure is described as follows:

M1.   Define the filter $b(z)$ using one of the zero phase, zero phase with gain compensation, zero phase with extended gain compensation, and zero phase with optimal gain compensation methods of Chapter 4 to minimise

$$|b(z)G_{cl}(z)|_{z=e^{j\omega T_s}} .$$

M2.   Define the set of frequencies to be used as

$$\Omega = \left\{ \omega_k = \frac{2\pi k}{NT_s}, k \in \{1, \cdots, n\} \mid \omega_1, \dots, \omega_l; \omega_{l+1}, \dots, \omega_m; \omega_{m+1}, \dots, \omega_n \right\},$$

where $\omega_1 = 2\pi/NT_s$ is the fundamental frequency of the periodic exogenous input and $\omega_k = 2\pi k/NT_s$ is the chosen bandwidth of repetitive control (limited by the bandwidth of the actuator used). Frequencies ω_{m+1} to ω_n are high frequencies where significant model uncertainty exists ($\omega_{m+1} > 10\omega_l$) and the intermediate frequencies ω_{l+1} to ω_m.

M3.   Choose a specific frequency value $\omega = \omega_i \in \Omega$, $\forall i \in \{1, \dots, n\}$ from set Ω in M2. $|W_S(\omega)|$, $|W_T(\omega)|$ and $|G(e^{j\omega T_s})|$ are known at this point.

M4.   Evaluate Δ_M by using (6.54) and select the active range of $\alpha \in [0, 2\pi]$ such that $\Delta_M \geq 0$. For all values of α in the active range:

   M4a.   Evaluate $|L|$ by using (6.53). Keep only the positive solutions.

   M4b.   Evaluate $L = |L|e^{j\alpha}$.

M4c. Solve for the repetitive controller filter $q_p(e^{j\omega T_s})$ at the chosen frequency ω by using

$$q_p(e^{j\omega T_s}) = \frac{L(z) - G(z)}{L(z) - G(z)[1 - b(z)]} z^{N-n_q}\Big|_{z=e^{j\omega T_s}}$$

M4d. Using the specific structure of $q_p(e^{j\omega T_s})$, back solve for the two chosen controller parameters within them. For example, $q_p(z)$ can be chosen as a multiplication of the second-order low-pass filters with unity d.c. gain

$$q_p(z) = \prod_{i=1}^{n} \left(\frac{1 + q_{1i} + q_{0i}}{1 + q_{3i} + q_{2i}}\right)\left(\frac{z^2 + q_{3i}z + q_{2i}}{z^2 + q_{1i}z + q_{0i}}\right). \tag{6.55}$$

There are four tuneable parameters for $n = 1$ in (6.55) which can be used to represent different types of controllers. These four tuneable parameters are q_{3i}, q_{2i}, q_{1i} and q_{0i}. If the performance of the filters which are utilised in the repetitive controller for $n = 1$ is unsatisfactory, n can be increased and new higher order filters can be synthesised. For the filter structure choice in (6.55), the back solution procedure uses

$$\text{Re}(q_p(j\omega)) = \kappa \frac{N_{re}(z)}{D(z)} \quad \text{and} \quad \text{Im}(q_p(j\omega)) = \kappa \frac{N_{im}(z)}{D(z)},$$

where

$$\kappa = \frac{1 + q_{1i} + q_{0i}}{1 + q_{3i} + q_{2i}},$$

$$N_{re}(z) = 2(q_{2i} + q_{0i})\cos^2 \omega T_s + (q_{1i} + q_{2i}q_{1i} + q_{3i} + q_{3i}q_{0i})\cos \omega T_s$$
$$+ (1 + q_{2i})(1 - q_{0i}) + q_{3i}q_{1i},$$

$$N_{im}(z) = (2(q_{0i} - q_{2i})\cos \omega T_s + q_{3i}q_{0i} + q_{1i} - q_{3i} - q_{2i}q_{1i})\sin \omega T_s,$$

$$D(z) = 4q_{0i}\cos^2 \omega T_s + 2(q_{0i} + q_{1i})\cos \omega T_s - 2q_{0i} + 1^2 + q_{1i}^2 + q_{0i}^2.$$

M5. The solution in step M4 leads to a closed curve which is plotted for solving two of the four parameters q_{0i} to q_{3i}. Plot the closed curve obtained in the chosen controller parameter space. Either the inside (drawn with a solid boundary) or outside (drawn with a dashed boundary) of this curve is a solution of (6.51) at the chosen frequency (see the ellipses in Figure 6.20(b), for example). The region obtained is the point condition solution in the chosen repetitive controller parameter plane at the frequency chosen in step M3.

M6. Go back to step M3 and repeat the procedure at a different frequency until all frequencies in set Ω are used.

M7. Plot the intersection of all point condition solutions for all frequencies in set Ω. This is the overall solution region for the robust performance requirement.

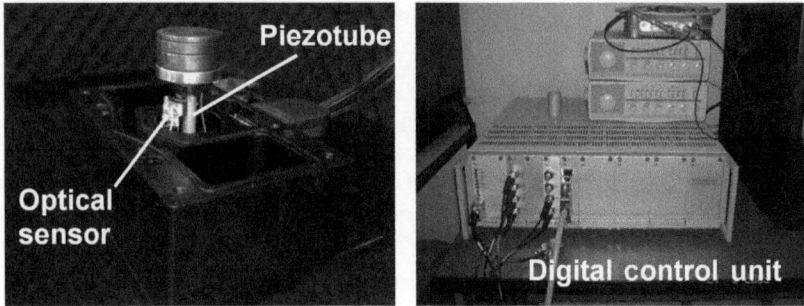

Figure 6.17   AFM setup and the piezotube used

Proposition 1. *The stability region of the repetitive controller can be computed by solving*

$$\|\mathscr{R}(\omega)\|_\infty \triangleq \sup_{\omega \in \Omega} \left(q(z) \left(1 - b(z) \frac{G(z)}{1 + G(z)} \right) \right) \Bigg|_{z=e^{j\omega T_s}}$$

for each frequency value of the set Ω. The repetitive control system will be stable if $\|\mathscr{R}(\omega)\|_\infty$ when evaluated for the frequencies in the set Ω is less than unity.

6.9   Case study: high-speed atomic force microscope scanner position control

In this part of Chapter 6, we use the model of a piezotube based AFM in our laboratory, cf. [57]. Piezotube scanners are extensively used in the high-speed AFM since they provide a higher positioning accuracy and larger bandwidth than other techniques. Additionally, they are easier to manufacture and to integrate into a microscope. Piezotubes consist of two parts: the inner and outer electrodes. The quartered outer electrode of a piezotube is used for the raster scan motion on the x–y plane, and the inner electrode is used for z-positioning (vertical motion).

Figure 6.17 shows the AFM setup and the piezotube, which are used in our laboratory. We, here, benefit from the piezotube scanner model as a numerical example to explain the methodology of the multi-objective parameter space approach for discrete-time repetitive control. The identified transfer function of the piezotube actuator in the x-direction (see Figure 6.18) is obtained as

$$G(z) = K \frac{-z^3 + 5.3902z^2 - 6.9626z + 3.5691}{z^4 - 3.1450z^3 + 4.3460z^2 - 2.9460z + 0.8777},$$

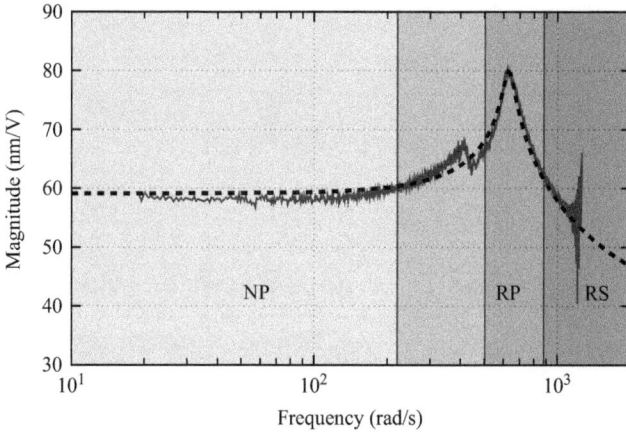

Figure 6.18   *The experimental (solid black) and identified (dashed black) Bode magnitude plots of piezotube actuator, mentioned in [57], are presented here. In this figure, NP stands for the nominal performance, RP represents the robust performance, and RS denotes the robust stability*

where $K = 120.2 \times 10^{-6}$ nm/V includes the power amplifier and sensor gains. The piezotube actuator model has one real NMP zero located at 3.8079. Using this fact, we separate the numerator of the transfer function $G(z)$ into two parts:

$$n_{mp}(z) = -120.2 \times 10^{-6}(z^2 - 1.5823z + 0.9373)$$

and

$$n_{nmp}(z) = z - 3.8079.$$

We now design the dynamic compensator $b_p(z)$ by using different approximate inverse filter design techniques described in Chapter 4. We firstly compute the zero phase (ZP) compensator as an approximate inverse filter for the piezotube actuator as

$$b_p(z) = \frac{z^{-2}d(z)}{n_{mp}(z)} \frac{-0.483z + 0.1268}{z}.$$

Next, we design the dynamic compensator $b_p(z)$ as an ZPGO compensator. For the optimisation problem in (4.41), the weighting function is chosen as

$$W(\theta) = \begin{cases} 1, & \text{if } \theta \leq \theta^* \\ 0, & \text{otherwise} \end{cases}$$

to have large gain at high frequencies above $\theta^* = {}^{4\pi}/s$rad. The weight was chosen by balancing the desired tracking bandwidth requirement (uniform weight up to θ^*) with the low gain requirement at high frequencies (large weight above θ^*) for robustness

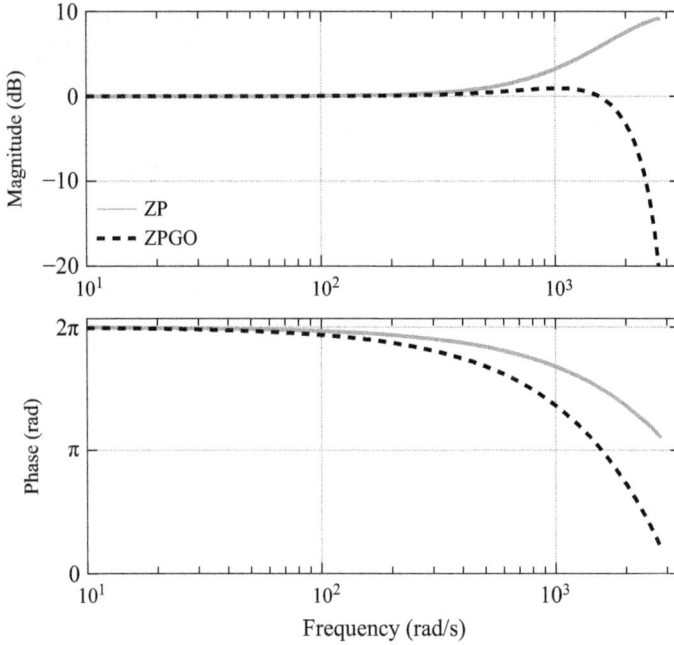

Figure 6.19   The Bode plots of the transfer function $b_p(z)G(z)$ with different kinds of dynamic compensators. In this plot, ZP denotes zero phase compensator, and ZPGO represents zero phase optimal gain compensator

of performance. Here, the optimal gain compensating zero is computed as $z_1^* = -1$ by solving the optimisation problem (4.41). The resulting compensator is given by

$$b_p(z) = \frac{z^{-3}d(z)}{n_{mp}(z)} \frac{-0.1207z^3 - 0.2098z^2 - 0.05733z + 0.03171}{z^2}.$$

Figure 6.19 illustrates a comparison of the magnitude (resp. phase) frequency responses of the ZP and ZPGO compensated systems. It also demonstrates the larger tracking bandwidth achieved by ZPGO compensation method. Therefore, we use ZPGO compensator as the dynamic compensator $b_p(z)$ in the parameter space computation.

Moreover, the time advance for the dynamic compensator $b(z)$ is chosen as

$$b(z) = b_p(z)z^{n_b} = b_p z^3 \tag{6.56}$$

to reduce the phase angle of $b(z)$ as much as possible. The low-pass filter $q(z)$ is chosen as

$$q(z) = q_p(z)z^{n_q} = \frac{(1 + q_{11} + q_{01})}{4} \frac{(z + 1)^2}{z^2 + q_{11}z + q_{01}} z^2. \tag{6.57}$$

Table 6.3   Desired sensitivity magnitude upper bounds at $NT_s = 0.1$ s

Frequency range	k	$f = k/NT_s$ (Hz)	W_S	W_T
Low (NP)	1	10	125	0
Low (NP)	2	20	50	0
Low (NP)	3	30	25	0
Low (NP)	4	40	15	0
Intermediate (RP)	8	80	3.5	0.15
Intermediate (RP)	10	100	2.5	0.10
Intermediate (RP)	12	120	2.0	0.10
Intermediate (RP)	14	140	1.6	0.05
High (RS)	16	160	0	0.9
High (RS)	18	180	0	0.9
High (RS)	20	200	0	0.9

The parameters of $q_p(z)$ given in (6.57) are chosen as $q_{11} = 2$ and $q_{01} = 1$ in the general form (6.55) to obtain unity d.c. gain. We add phase advance to this low-pass filter for phase cancellation, and we thus aim at reducing the steady-state error of the closed-loop control system. The region in the $q_{01} - q_{11}$ controller parameter space is computed for three cases which are, respectively, nominal performance (NP) at low frequencies ($W_T = 0$), robust performance (RP) at intermediate frequencies and robust stability (RS) at high frequencies ($W_S = 0$), as shown in Figure 6.18.

The sensitivity constraints are specified at a set of discrete frequencies. The periodic input command of the high-speed AFM scanner has a period of 100 ms. The specific numerical values of the chosen weights used in the computation of the controller parameters are tabulated in Table 6.3. Figure 6.20 (a) shows the intersection of the stability region of the repetitive controller (6.57) without any delay element (i.e., described by the triangular region) and the sufficient stability condition $\mathcal{R}(\omega) < 1$ for all $\omega \in \Omega$, whereas Figure 6.20(b) demonstrates the overall region calculated for nominal performance, robust performance and stability robustness.

The mathematical model of the piezotube actuator used as the application example here is valid for frequencies up to 160 Hz. A uniform weight for the robust stability requirement for frequencies above this value is chosen here. The relative multiplicative error $|(G - G_n)/G_n|$ has to be below the weight specified in the stability robustness considerations given in Figure 6.18. The intersection of the regions, which are calculated for the nominal performance, the robust performance and the robust stability requirements, in the $q_{01} - q_{11}$ controller parameter space is filled with gray colour. The solution procedure is concluded by choosing a point in the controller parameter plane given in Figure 6.20 (b). The solution point within this region is chosen arbitrarily in this example and is marked with a cross in Figure 6.20(b). The simulation result for a triangular wave input with the period 10 Hz is displayed in Figure 6.21. This result shows the effectiveness of the repetitive controller in decreasing the steady-state error while tracking a periodic input signal.

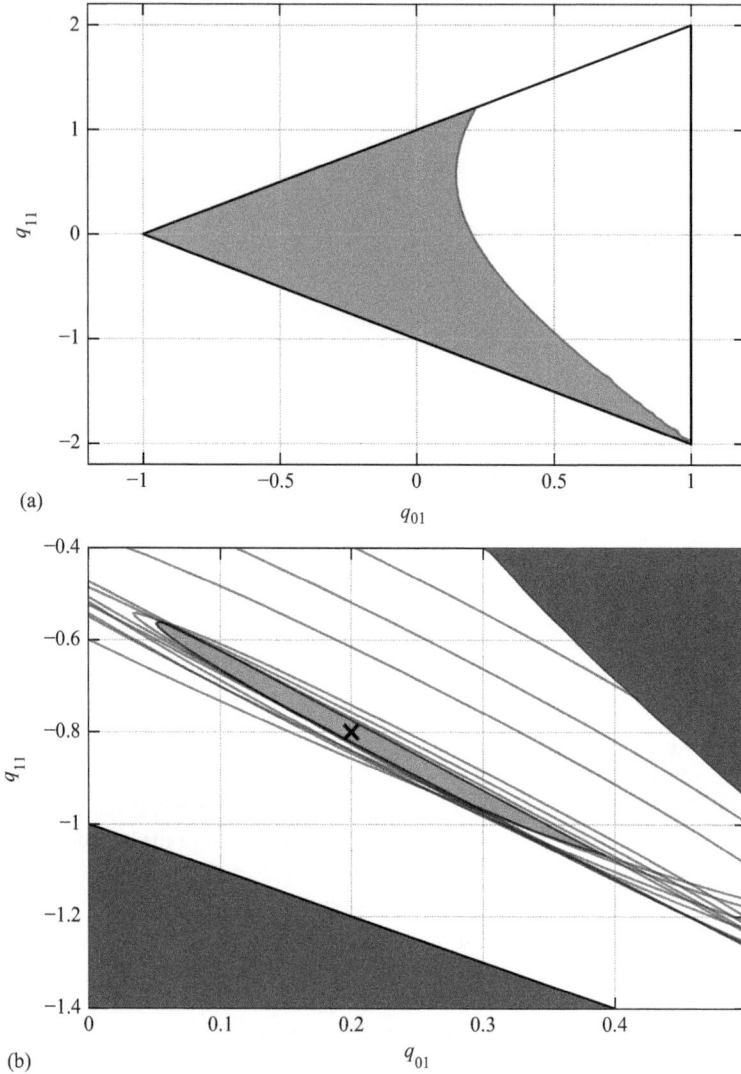

(a)

(b)

Figure 6.20   Parameter space representation of the repetitive control system:
(a) stability region (i.e., $\mathcal{R}(\omega) < 1$), (b) intersection of the stability
and the performance regions

6.10   Chapter summary and concluding remarks

This chapter presented repetitive control, which is a well-known servo-control technique used for accurately tracking a periodic reference signal, or for rejecting a periodic disturbance with a known period. The main element of a repetitive controller

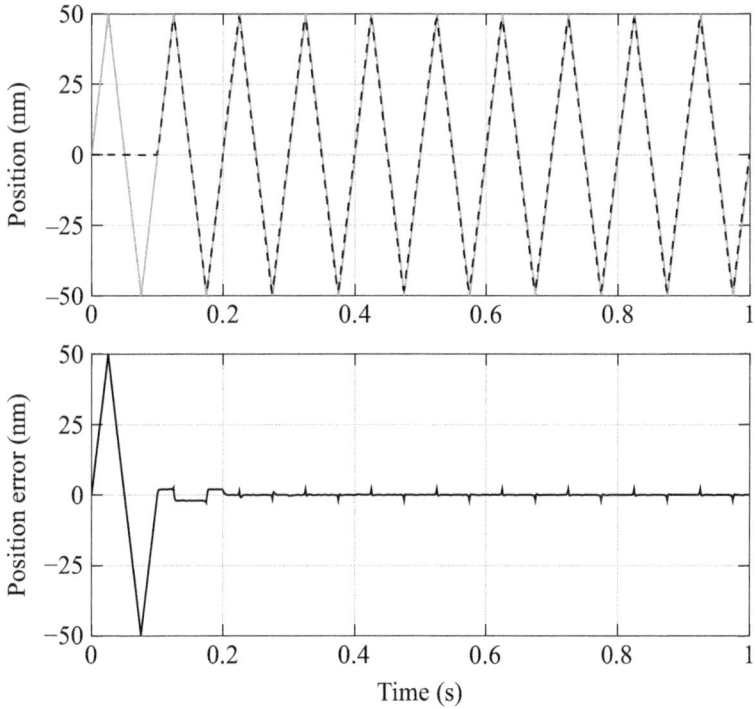

Figure 6.21   Simulation results for triangular wave input at 10 *Hz*

is a generator of periodic signals formed by a positive feedback loop that contains a time delay element. The delay in the periodic signal generator is equal to the known period of the periodic reference (or disturbance) signal. This chapter introduced a continuous-time repetitive controller design procedure, based on a multi-objective parameter space approach, to satisfy a desired robust performance specification at the fundamental frequency and selected harmonics. Additionally, the discrete-time extension of this repetitive controller design approach was presented. Lastly, this chapter demonstrated the effectiveness of the proposed design technique via simulation studies based on the position control of a high-speed AFM and the Quanser QUBE™ servo system.

References

[1]   B. A. Francis and W. A. Wonham, "The internal model principle for linear multivariable regulators," *Applied Mathematics and Optimization*, vol. 2, no. 2, pp. 170–194, May 1975.

[2]   T. Inoue, S. Iwai, and M. Nakano, "High accuracy control of a proton synchrotron magnet power supply," in *Proceedings of Eighth IFAC World Congress*, 1981.

[3] H. Fujimoto, F. Kawakami, and S. Kondo, "Switching based repetitive con-
     trol for hard disk drives: Experiments on RIDR and RPTC methods," in
     Proceedings of the IEEE International Conference on Control Applications,
     2004.
[4] J. H. Moon, M. N. Lee, and M. Chung, "Repetitive control for the track-
     following servo system of an optical disk drive," *IEEE Transactions on Control
     Systems Technology*, vol. 6, no. 5, pp. 663–667, 1998.
[5] T.-C. Tsao and M. Tomizuka, "Robust adaptive and repetitive digital tracking
     control and application to a hydraulic servo for noncircular machining," *ASME
     Journal of Dynamic Systems, Measurement and Control*, vol. 116, no. 1, pp.
     24–32, March 1994.
[6] T. Omata, S. Hara, and M. Nakano, "Nonlinear repetitive control with applica-
     tion to trajectory control of manipulators," *Journal of Robotic Systems*, vol. 4,
     no. 5, pp. 631–652, Oct. 1987.
[7] N. Sadegh, R. Horowitz, W. W. Kao, and M. Tomizuka, "A unified approach
     to the design of adaptive and repetitive controllers for robotic manipulators,"
     ASME Journal of Dynamical Systems, Measurement and Control, vol. 112,
     no. 4, pp. 618–629, Dec. 1990.
[8] K. Kaneko and R. Horowitz, "Repetitive and adaptive control of robot
     manipulators with velocity estimation," *IEEE Transactions on Robotics and
     Automation*, vol. 13, no. 2, pp. 204–217, April 1997.
[9] J. Kasac, B. Novakovic, D. Majetic, and D. Brezak, "Passive finite-dimensional
     repetitive control of robot manipulators," *IEEE Transactions on Control
     Systems Technology*, vol. 16, no. 3, pp. 570–576, May 2008.
[10] F. Kobayashi, S. Hara, and H. Tanaka, "Reduction of motor speed fluctua-
     tion using repetitive control," in *Proceedings of the 29th IEEE Conference on
     Decision and Control*, 1990.
[11] T. Katayama, M. Ogawa, and M. Nagasawa, "High-precision tracking control
     system for digital video disk players," *IEEE Trans. Consum. Electron.*, vol. 41,
     no. 2, May 1995.
[12] R. F. Fung, J. S. Huang, and C. C. Chien, "Design and application of a contin-
     uous repetitive controller for rotating mechanisms," *International Journal of
     Mechanical Sciences*, vol. 42, no. 9, pp. 1805–1819, Sept. 2000.
[13] J. Wang and T.-C. Tsao, "Repetitive control of linear time varying systems
     with application to electronic cam motion control," in *Proceeding of the 2004
     American Control Conference*, 2004.
[14] F. R. Shaw and K. Srinivasan, "Discrete-time repetitive control systems
     design using the regeneration spectrum," *ASME Journal of Dynamic Systems,
     Measurement and Control*, vol. 115, no. 2A, pp. 228–237, June 1993.
[15] S. S. Garimella and K. Srinivasan, "Application of repetitive control to
     eccentricity compensation in rolling," *ASME Journal of Dynamical Systems,
     Measurement and Control*, vol. 118, no. 4, pp. 657–664, Dec. 1996.
[16] G. Hillerström, "Adaptive suppression of vibrations-a repetitive control
     approach," *IEEE Transactions on Control Systems Technology*, vol. 4, no. 1,
     pp. 72–78, Jan. 1996.

[17] S. Tinghsu, S. Hattori, M. Ishida, and T. Hori, "Suppression control method for torque vibration of AC motor utilizing repetitive controller with Fourier transform," *IEEE Transactions on Industry Application*, vol. 38, no. 5, pp. 1316–1325, Sept./Oct. 2002.

[18] G. Pipeleers, B. Demeulenaere, F. Al-Bender, J. De Schutter, and J. Swevers, "Optimal performance tradeoffs in repetitive control: Experimental validation on an active air bearing setup," *IEEE Transactions on Control Systems Technology*, vol. 17, no. 4, pp. 970–979, July 2009.

[19] Y. Li, G. T.-C. Chiu, and L. G. Mongeau, "Dual-driver standing wave tube: Acoustic impedance matching with robust repetitive control," *IEEE Transactions on Control Systems Technology*, vol. 12, no. 6, pp. 869–880, Nov. 2004.

[20] C. Rech, H. Pinheiro, and H. Grundling, "Comparison of digital control techniques with repetitive integral action for low cost PWM inverters," *IEEE Transactions on Power Electronics*, vol. 18, no. 1, pp. 401–410, Jan. 2003.

[21] K. Zhang, Y. Kang, and J. Xiong, "Direct repetitive control of SPWM inverter for UPS purpose," *IEEE Transactions on Power Electronics*, vol. 18, no. 3, pp. 784–792, May 2003.

[22] K. Zhou and D. Wang, "Digital repetitive controlled three-phase PWM rectifier," *IEEE Transactions on Power Electronics*, vol. 18, no. 1, pp. 309–316, March 2003.

[23] G. S. Choi, Y. A. Lim, and G. H. Choi, "Tracking position control of piezoelectric actuators for periodic reference inputs," *Mechatronics*, vol. 12, no. 5, pp. 669–684, June 2002.

[24] G. H. Choi, H. O. Jong, and G. S. Choi, "Repetitive tracking control of a course-fine actuator," in *Proceedings of IEEE/ASME International Conference Advanced Intelligent Mechatronics*, 1999.

[25] U. Aridoğan, Y. Shan, and K. K. Leang, "Design and analysis of discrete-time repetitive control for scanning probe microscopy," *ASME Journal of Dynamic Systems, Measurement and Control*, vol. 131, no. 6, p. 061103, Oct. 2009.

[26] R. J. E. Merry, M. J. C. Ronde, R. van de Molengraft, K. R. Koops, and M. Steinbuch, "Directional repetitive control of a metrological AFM," *IEEE Transactions on Control Systems Technology*, vol. 19, no. 6, pp. 1622–1629, Nov. 2011.

[27] S. Necipoğlu, S. A. Cebeci, C. Başdoğan, Y. E. Has, and L. Güvenç, "Repetitive control of an XYZ piezo-stage for faster nano-scanning: Numerical simulations and experiment," *Mechatronics, accepted for publication*, vol. 21, no. 6, pp. 1098–1107, Sept. 2011.

[28] S. Necipoğlu, S. A. Cebeci, Y. E. Has, L. Güvenç, and C. Başdoğan, "A robust repetitive controller for fast AFM imaging," *IEEE Transactions on Nanotechnology*, vol. 10, no. 5, pp. 1074–1082, Jan. 2011.

[29] S. Hara, Y. Yamamoto, T. Omata, and M. Nakano, "Repetitive control system: A new type servo system for periodic exogenous signals," *IEEE Transactions on Automatic Control*, vol. 33, no. 7, pp. 659–668, July 1988.

[30]   Y. Yamamoto and S. Hara, "Relationships between internal and external stability for infinite-dimensional systems with applications to a servo problem," *IEEE Transactions on Automatic Control*, vol. 33, no. 11, pp. 1044–1052, Nov. 1988.

[31]   K. Srinivasan and F. Shaw, "Analysis and design of repetitive control systems using the regeneration spectrum," *ASME Journal of Dynamic Systems, Measurement and Control*, vol. 113, no. 2, pp. 216–222, June 1991.

[32]   M. Tomizuka, T. C. Tsao, and K. K. Chew, "Analysis and synthesis of discrete-time repetitive controllers," *ASME Journal of Dynamical Systems, Measurement and Control*, vol. 111, no. 3, pp. 353–358, Sept. 1989.

[33]   K. Srinivasan, H. Özbay, and I. S. Jung, "A design procedure for repetitive control systems," in *Proceedings of ASME International Mechanical Engineer's Congress and Exposition*, 1995.

[34]   H. Özbay, "H_∞ optimal controller design for a class of distributed parameter space," *International Journal of Control*, vol. 58, no. 4, pp. 739–782, Oct. 1993.

[35]   T. Peery and H. Özbay, "H_∞ optimal repetitive controller design for stable plants," *ASME Journal of Dynamical Systems, Measurement and Control*, vol. 119, no. 3, pp. 541–547, 1997.

[36]   C. L. Roh and M. J. Chung, "Design of repetitive control system for an uncertain plant," *Electronics Letter*, vol. 31, no. 22, pp. 1959–1960, Oct. 1995.

[37]   G. Weiss and M. Häfele, "Repetitive control of MIMO systems using H_∞ design," *Automatica*, vol. 35, no. 7, pp. 1185–1199, July 1999.

[38]   G. Weiss, Q.-C. Zhong, T. C. Green, and J. Liang, "H_∞ repetitive control of DC-AC converters in microgrids," *IEEE Transactions on Power Electronics*, vol. 19, no. 1, pp. 219–230, Jan. 2004.

[39]   Q.-C. Zhong, J. Liang, G. Weiss, C. Feng, and T. C. Green, "H_∞ control of the neutral point in four-wire three-phase DC-AC converters," *IEEE Transactions on Industrial Electronics*, vol. 53, no. 5, pp. 1594–1602, Oct. 2006.

[40]   L. Güvenç, "Stability and performance robustness analysis of repetitive control systems," *ASME Journal of Dynamic Systems, Measurement and Control*, vol. 118, no. 3, pp. 593–597, Sept. 1996.

[41]   J. Li and T. C. Tsao, "Robust performance repetitive control design using structured singular values," in *Proceedings of ASME International Mechanical Engineer's Congress and Exposition*, 1998.

[42]   B. Aksun-Güvenç, "Applied robust motion control," Ph.D. dissertation, Istanbul Technical University, Istanbul, Turkey, 2001.

[43]   B. A. Güvenç and L. Güvenç, "Robust repetitive controller design in parameter space," *ASME Journal of Dynamical Systems, Measurement and Control*, vol. 128, no. 2, pp. 406–413, March 2006.

[44]   B. Demirel and L. Güvenç, "Parameter space design of repetitive controllers for satisfying a robust preformance requirement," *IEEE Transactions on Automatic Control*, vol. 55, no. 8, pp. 1893–1899, Aug. 2010.

[45]   T.-Y. Doh and M. J. Chung, "Repetitive control design for linear systems with time-varying uncertainties," *IEEE Control Theory and Applications*, vol. 123, no. 4, pp. 427–432, 2003.

[46] S. Yi, P. W. Nelson, and A. G. Ulsoy, *Time-Delay Systems Analysis and Control Using the Lambert W Function*, ser. Technology & Engineering. World Scientific, 2010.

[47] E. Fridman, *Introduction to Time-Delay Systems: Analysis and Control*, ser. Systems & Control: Foundations & Applications. Springer, 2014.

[48] T. Omata, S. Hara, and M. Nakano, "Synthesis of repetitive control systems and its applications," in *Proceedings of the 24th IEEE Conference on Decision and Control*, 1985.

[49] N. Sadegh, "Synthesis and stability analysis of repetitive controllers," in *Proceedings of American Control Conference*, 1991.

[50] S. Hara and Y. Yamamoto, "Stability of repetitive control systems," in *Proceeding of 24th IEEE Conference on Decision and Control*, 1985.

[51] G. J. Silva, A. Datta, and S. P. Bhattachaiyya, *PID Controllers for Time-Delay Systems*, ser. Control Engineering. Birkhäuser Basel, 2005.

[52] K. Srinivasan and C. L. Nachtigal, "Analysis and design of machine tool chatter control systems using the regeneration spectrum," *ASME Journal of Dynamic Systems, Measurement and Control*, vol. 100, no. 3, pp. 191–200, Sept. 1978.

[53] H. L. Broberg and R. G. Molyet, "A new approach to phase cancellation in repetitive control," in *Proceedings of Industry Applications Society Annual Meeting*, 1994.

[54] J. Ackermann, P. Blue, T. Bünte, L. Güvenç, D. Kaesbauer, M. Kordt, M. Mühler, and D. Odhental, *Robust Control: The Parameter Space Approach*. Springer Verlag: London, UK, 2002.

[55] G. Schitter, K. J. Åström, B. E. DeMartini, P. J. Thurner, K. L. Turner, and P. K. Hansma, "Design and modeling of a high-speed AFM-scanner," *IEEE Transactions on Control Systems Technology*, vol. 15, no. 5, pp. 906–915, Sept. 2007.

[56] B. Demirel, "Interactive computer-aided controller design for mechatronic systems," M. Sc. Thesis, Istanbul Technical University, Istanbul, Turkey, Sept. 2009.

[57] B. Aksun-Güvenç, S. Necipoğlu, B. Demirel, and L. Güvenç, *Robust Control of Atomic Force Microscopy*, ser. Mechatronics. London: ISTE/Wiley, March 2011, ch. 3.

Chapter 7

Summary and conclusions

Several different methods that have been successfully applied in the past for controlling mechatronic systems are presented in this book. In this final chapter of the book, the main findings are first summarised and then followed by conclusions.

7.1 Summary

The control methods treated in this book were conventional control methods (phase lead, lag, PID), input shaping feedforward controllers with preview, disturbance observer compensation and repetitive control. Along with the conventional approaches, the parameter space method of robust control was also introduced and applied to most of the control methods treated in this book. The parameter space methods that were introduced and used were analysis and design for Hurwitz stability, $\mathscr{D}$-stability, phase and gain margin, and mixed sensitivity bounds. More conservative Nyquist stability based robustness conditions were also introduced and used for unstructured model uncertainty.

As a pre-requisite to the following sections, the parameter space approach to robustness analysis and robust controller design were introduced and treated in Chapter 2. Phase lead, lag, lag-lead and PID controller design were treated in Chapter 3, using analytical design equations for a specified static error constant and a specified phase margin at a specified gain crossover frequency. It was seen that the analytical equations obtained could be used for a phase margin bound design like that of Chapter 2. The focus of the book switched to discrete-time plants under feedback control in Chapter 4, which started with the occurrence of non-minimum phase zeros that may arise during the discretisation process, and followed with a characterisation on non-minimum phase zeros. Several input preview based zero phase attaining methods were proposed as approximate inverse filters for inverting the feedback controlled closed-loop system. It was noted that these approximate inverses could also be used in later chapters with or without the preview. Disturbance observer compensation was presented in Chapter 5 as an excellent disturbance rejection system with very good desired model regulation features. A decoupling extension and a communication disturbance observer were also formulated and presented to handle square MIMO plants and plants with variable network time delays, respectively. Systems with periodic reference or disturbance inputs were treated in Chapter 6, with repetitive control being introduced as the solution. It was seen that the repetitive controller did not need knowledge of the shape of the periodic input, with knowledge of its period being sufficient. Mixed sensitivity parameter space design was introduced in both Chapters 5 and 6 for tuning of chosen controller parameters.

All chapters used textbook type examples and more realistic case studies to illustrate the concepts. The case studies ranged from road vehicle yaw stability control and automated path following, to decoupling control of piezotube actuators in an AFM. The Quanser QUBE™ servo rotational speed and position control loops were also used in some of the case studies. The MATLAB® code for most of the examples have been made available in the book's website.

7.2 Conclusions

This book is the result of close to two decades of work of the authors on modelling, simulating and controlling different mechatronic systems from the motion control, automotive control and micro and nano-mechanical systems control areas. The methods presented in the book have all been tested by the authors and a very large group of researchers, who have produced practically implementable controllers with highly successful results. The approach that is recommended in this book is to first start with a conventional control method which may then be cascaded with a feedforward controller if the input is known or can be measured with a preview, to add a disturbance observer if unknown disturbances are to be rejected and if regulation of the uncertain plant about a nominal model is desired, and to add a repetitive controller to take care of any periodic inputs of fixed and known period. While a lot of results on the application of parameter space methods and discrete-time and MIMO extensions were presented, more work still needs to be carried out and presented in a future edition of the book.

There are also other control methods that are used in controlling mechatronic systems, or show potential advantages but have not been exploited enough. Two methods that have been used successfully in the past for controlling mechatronic systems are iterative learning control and model predictive control. They were not part of the present book due to space limitations and the fact that parameter space methods have not been applied to them. Fractional control systems, in which non-integer fractions of the constant, integral and derivative terms are used, also show some potential advantages, especially in combination with parameter space methods for design and analysis. The reader is referred to the existing literature for more details of iterative learning control, model predictive control and fractional control systems.

Appendix A
Rapid control prototyping
and hardware-in-the-loop simulation

Hardware-in-the-loop simulation and rapid controller prototyping are two techniques that are used frequently in the V-cycle design process of control systems for mechatronic products and are, thus, briefly reviewed here.

The V-cycle design process for control system design is a sequence of model-in-the-loop, software-in-the-loop, hardware-in-the-loop and real-world experiments down and up the V-cycle; see Figure A.1.

Model-in-the-loop simulation or offline simulation are the Simulink® (or other modelling program) simulations that have been used throughout this book.

Rapid controller prototyping is the generation of C (or other real time implementable) code of a controller from its Simulink diagram, linking it and downloading it to a target processor (hardware) like an electronic control unit. The control code generated by rapid controller prototyping can be used in software-in-the-loop,

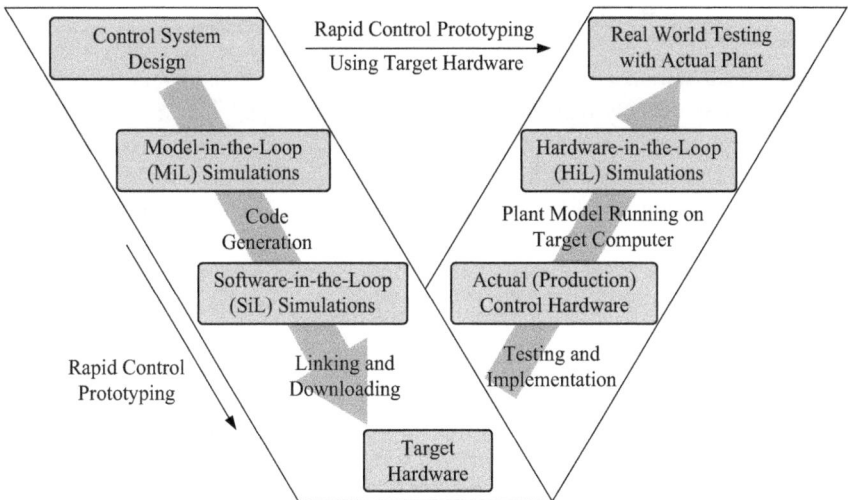

Figure A.1   V-cycle of control system design for mechatronic product

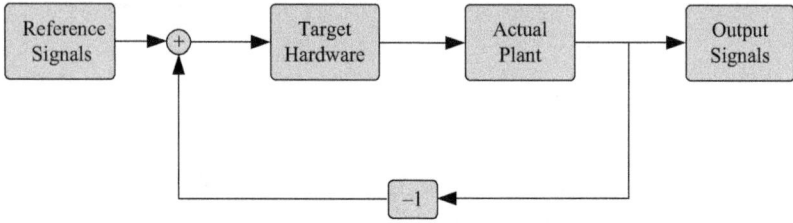

Figure A.2   Rapid control prototyping

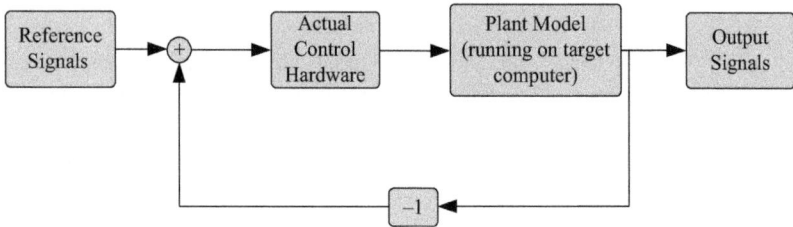

Figure A.3   Hardware-in-the-Loop simulation

hardware-in-the-loop and real-world testing. Figure A.2 indicates a real-world testing application of rapid control prototyping.

In software-in-the-loop testing, the actual control code that is usually generated by rapid controller prototyping is used in an offline simulation. As the actual control code is being used, the code can be checked for bugs and errors.

Timing and real-time implementation problems cannot be checked in software-in-the-loop simulation. For that purpose, a fast and powerful computer (named target computer) runs a highly realistic model of the mechatronic plant in real-time. The actual control computation hardware, which is usually a generic electronic control unit, is connected to this target computer running the plant model with exactly the same input and output interfaces as in the real-world mechatronic product. The control code and control hardware are fooled into believing that they are connected to the actual plant. This is called hardware-in-the-loop simulation and is a very powerful tool, as experimental conditions that are very difficult to generate can easily be tested in a safe lab environment; see Figure A.3.

The last step in the V-cycle design is real-world testing where rapid controller prototyping is used again (like in hardware-in-the-loop simulation) to make controller changes on the fly. The V-cycle design procedure is a very powerful, efficient and proven approach to control system design and development for a mechatronic system.

Index